An Approach towards Generic Coastal Geomorphological Modelling with Applications

An Approach towards Generic Coastal Geomorphological Modelling with Applications

DISSERTATION
Submitted in fulfilment of the requirement of
the Board of Doctorates of Delft University of Technology
and of the Academic Board of the UNESCO-IHE Institute for Water Education
for the Degree of DOCTOR
to be defended in public
on Tuesday, 05 June 2012 at 12:30 hours
in Delft, The Netherlands

by

Qinghua, YE
born in Hubei, China
Bachelor of Science, Zhejiang University (former Hangzhou University), Hangzhou, China
Master of Science, Nanjing Hydraulic Research Institute, Nanjing, China
Master of Science, UNESCO-IHE, Delft, The Netherlands

This dissertation has been approved by the supervisor:
Prof.dr.ir. J.A. Roelvink

Composition of Doctoral Committee:

Chairman	Rector Magnificus Delft University of Technology
Vice-Chairman	Rector UNESCO-IHE
Prof.dr.ir. J.A. Roelvink	Delft University of Technology / UNESCO-IHE, supervisor
Prof.dr.ir. J.H.W. Lee	University of Hongkong /
	HongKong University of Science and Technology
Prof.dr. P.M.J. Herman	NIOZ / Radboud University Nijmegen
Prof.dr.ir. A.E. Mynett	UNESCO-IHE / Delft University of Technology
Prof.dr.ir. G.S. Stelling	Delft University of Technology
Dr.ir. H.R.A. Jagers	Deltares
Prof.dr.ir. M.J.F. Stive	Delft University of Technology, reserve member

CRC Press/Balkema is an imprint of the Taylor & Francis Group, an informa business

Published by:
CRC Press/Balkema
PO Box 447, 2300 AK Leiden, the Netherlands
e-mail: Pub.NL@taylorandfrancis.com
www.crcpress.com - www.taylorandfrancis.co.uk - www.ba.balkema.nl

ISBN 978-0-415-64160-9 (Taylor & Francis Group)

Ministerie van Infrastructuur en Milieu

Deltares and the Netherlands Ministry of Infrastructure and Environment are gratefully acknowledged for their financial support to this study.

Abstract

Numerical modelling approach is widely used to study geomorphology in coastal systems nowadays. The earlier generations of numerical model domain were usually discretized to structured or unstructured grids in one, two or three dimensions. Different discretization approaches had their advantages and disadvantages, for example, the matrix systems from structured grid generally had better numerical properties, thus might be easier to be solved; while the unstructured grid may provide more flexibility to arrange model domain for complicate geometry which is usually the case in coastal zones. Discretization using unstructured grids may avoid the "staircase problem" which commonly occurs in structured grids.

In this study, a generic morphological model was developed which is flexible for various types of computational grids including both structured and unstructured grids. The model is be able to coupled online with a number of hydrodynamic models, such as the 1D/2D flow model Sobek, the 2D/3D structured-grid flow model Delft3D-FLOW, and the 2D/3D unstructured-grid flow model D-Flow FM.

To achieve the flexibility of the generic model, an open structure model framework and a number of adaptive algorithms were developed. For sediment mass conservation, a staggered grid Finite Volume method was used in the bed level updating equation. The mass transport was solved by a upwind and a second order center scheme, and a total variation diminishing (TVD) scheme was applied to avoid numerical oscillations, by indicating that no new maxima and minima are generated higher or lower than those that were already in the solution. To estimate velocity vector at the grid centre, a flexible velocity integration algorithm was developed for both structured and unstructured grids. Remediate approaches for violation of total mass conservation due to morphological updating were proposed.

The model framework was designed based on the Object Oriented Programming (OOP) concept using the open processes library of Delft3D-WAQ. Under this framework, sediment (sand and mud) and vegetation were presented as substances in the substance library, while the corresponding morphodynamic processes, vegetation population dynamic processes and interactions were prescribed as processes in the process library. The morphodynamic processes ranging from particle scale to geomorphology scale included: 1) sediment (sand/mud) entrainment, transport, (hindered) settling, deposition; 2) acceleration and deceleration flow; 3) bed slope effects; 4) sediment exchange with bed book keeping system, bed form and the corresponding bed roughness. Vegetation population dynamic processes, on the other hand, included vegetation growth, spatial extension, seed dispersion, competition among species. Their effects on flow, wave and sediment transport were estimated together with the feedback of the ambient environment impacts on vegetation dynamics. The functions of the morphological model had been validated against analytical solutions, flume experiments and other morphological models. Validations of ecological processes had been carried out against the field measured spatial variation, distribution and abundance of two macrophyte species in Lake Veluwe, NL (from 1994 to 1997). The flexibility of open model structure allows more physical and biological processes to be integrated in future, such as wave related processes and aeolian transport.

Capability of the model had been explored by means of three engineering problems in China and the US. Firstly, it was used to study the bio-geomorphology in the restoration of a salt marsh in San Francisco Bay. The interaction of morphodynamic processes and ecological processes were considered simultaneously. This study focused on the channel network development with dynamic vegetation effects in tidal salt marsh. Traditional theories would regard that the presence of vegetation reduces sediment erosion (lower source). Nevertheless, the dynamics of vegetation patches affected the hydrodynamics because that more streamwise momentum was absorbed by the vegetations via drag forces (lower sediment transport capacity). The overall sedimentation pattern and channel development were the combination of these two effects. Thus it was found that the channel patterns depend not only on the channel width, basin area, but also on vegetation abundance, vegetation information updating frequency, colonization patterns and other ecological properties. This innovative bio-geomorphological modeling approach offers this possibility to study the dynamic interaction between vegetation population dynamics and morphology change.

Secondly, the model was applied to the study of the salt marsh restoration project in Nisqually estuary, south of Puget Sound, US. This study focuses on: 1) sedimentation patterns on the salt marsh and in the river channels; 2) vegetation patterns after the dam removal. With vegetation effect, sedimentation decreased in the river and increased on the tidal flat area where the vegetation grow and spread. In the previous embanked area, the sedimentation volume with vegetation effects was significantly higher than the volume without vegetation, which in turn helps the vegetation spread. Also for the existing salt marsh area, after the vegetation had been built up, the erosion volume at that area was much less than that without vegetation. The distribution pattern of vegetation species was also qualitatively predicted. The model results helped to explain and to predict the morphological change with effect of vegetation population dynamics in salt marshes.

Lastly, it was used to study the large scale morphodynamics of the shoreface connected radial sand ridges located in southern Yellow Sea, China. This study focused on the generation and long-term evolution trend of this unique large morphological feature. The modeling study helped to increase the understanding of the large scale tidal sand bank behavior. It was found that: 1) the radial tidal wave pattern may not be generated only by the two joint tidal waves, i.e. the tidal wave reflected by Shandong Peninsula and the progressive wave from Eastern China Sea, where the mild bed slope of the continental shelf is also essential. 2) Morphological modelling results show that the radial tide wave does not necessarily form the huge sand ridge system in radial shape.

In summary, a generic morphological model was developed, which had the flexibility to couple with various hydrodynamic models discretized in both structured and unstructured grids. The open structure of model frame and adaptive numerical algorithms allows the flexibility of the model. By integrating with ecological processes it was extended to become a bio-geomorphological model, which further showed the models extendibility.

The model had been validated for a wide range of test cases and had been applied to real practical problems. The results showed that this study improved our capability to analyse and to predict the coastal system with the increased understanding on coastal morphodynamic processes and ecological processes and their interactions. This generic geomorphological model can be used as a multidiscipline research tool for both morphologists and ecologists.

Samenvatting

De numerieke modelmatige methode wordt tegenwoordig alom gebruikt voor de geomorfologische bestudering van kustgebieden. Voorgaande numerieke modeldomeinen werden meestal gediscretiseerd als gestructureerde en ongestructureerde roosters in n, twee of drie dimensies. Deze discretisatiemethoden hadden elk afzonderlijk hun eigen voor- en nadelen; bijvoorbeeld, het matrixsysteem van een gestructureerde rooster heeft over het algemeen betere numerieke eigenschappen, en zou daardoor dus eenvoudiger zijn op te lossen, terwijl het ongestructureerde rooster meer flexibiliteit biedt voor de samenstelling van een modeldomein met een complexe geometrie, hetgeen in kustgebieden meestal het geval is. Discretisatie met behulp van ongestructureerde roosters kan het "trapprobleem" voorkomen dat zich doorgaans in gestructureerde roosters voordoet.

Deze studie richt zich op de ontwikkeling van een generiek morfologisch model dat kan worden aangepast aan diverse types computationele roosters, zowel gestructureerde als ongestructureerde. Een aantal hydrodynamische modellen, zoals het 1D/2D stromingsmodel Sobek, het 2D/3D gestructureerde rooster stromingsmodel Delft3D-FLOW en het 2D/3D ongestructureerde rooster stromingsmodel D-Flow FM kunnen online met het model worden gekoppeld.

Teneinde de flexibiliteit van het generieke model te realiseren is een modelraamwerk met open structuur en een aantal adaptieve algoritmes ontwikkeld. Voor het behoud van massa van sediment wordt in het vergelijkingen voor updates van de bodemhoogte gebruik gemaakt van een Eindige Volume methode met verspringend raster. Het massatransportprobleem kan worden opgelost met een upwind en een central numerieke schema, en door toepassing van een total variation diminishing scheme - TVD schema - ter vermijding van numerieke schommelingen door aan te geven dat er geen nieuwe maxima en minima worden gegenereerd die hoger of lager zijn dan die welke reeds in de oplossing voorkomen. Om de snelheidsvector op een centraal roosterpunt in te schatten is een algoritme ontwikkeld om snelheid flexibel te kunnen integreren voor zowel gestructureerde als ongestructureerde roosters. Ook worden voorstellen gedaan voor herstelmethoden ingeval van verstoring van het totale behoud van massa tengevolge van morfologische updates.

Het raamwerkmodel is aan de hand van de Open Processen Bibliotheek van Delft3D-WAQ ontworpen op basis van het objectgeoriënteerde programmeerconcept (OOP). Binnen dit raamwerk zijn sediment (zand en slib) en vegetatie als stoffen opgenomen in de stofbibliotheek, terwijl de daarmee corresponderende morfodynamische processen, de dynamische processen en interacties van vegetatiepopulatie, als processen zijn voorgeschreven in de procesbibliotheek. Deze morfodynamische processen die zich voordoen vanaf de schaal van een korreldeeltje tot de geomorfologische schaal omvatten: 1) meevoering van sediment (zand/slib), transport, (belemmerde) bezinking, depositie; 2) versnelling en -vertraging; 3) bodemhellingseffecten; 4) sedimentuitwisseling met bodem-boekhoudsysteem, de vorm van de bedding en de daarmee corresponderende ruwheid. En de dynamische processen van vegetatiepopulatie omvatten: vegetatiegroei, ruimtelijke uitbreiding, zaadverspreiding en concurrentie tussen soorten. Er is een inschatting gemaakt van de effecten daarvan op doorstroming, golf- en sedimenttransport samen met de feedback van het omgevingseffect op de vegetatiedynamiek. De functies van het morfologische model zijn gevalideerd tegenover analytische oplossingen, gootexperimenten en andere morfologische modellen. Een validatie van diverse ecologische processen is uitgevoerd tegenover de ruimtelijke variatie, spreiding en abondantie van twee macrofytensoorten welke van 1994 tot 1997 in het Veluwemeer, Nederland, in het veld (in situ) zijn gemeten. De flexibiliteit van dit model open structuur maakt het in de toekomst mogelijk om meer fysische en biologische processen, zoals golfgerelateerde processen en eolisch transport, (in het model) te integreren.

Mogelijkheden van het model zijn onderzocht door middel van drie probleemgevallen in China en de VS. Ten eerste is het gebruikt om de bio-geomorfologische processen bij het herstel van een kwelder in de Baai van San Francisco te bestuderen. De interactie van morfodynamische processen met ecologische processen is tegelijk bekeken. Deze studie richt zich op de netwerkontwikkeling van de geul met dynamische vegetatie effecten in een aan getij onderhevige kwelder. Volgens de meer traditionele theorieën zou de aanwezigheid van vegetatie leiden tot vermindering van sediment erosie (lagere bron). Desondanks is de dynamiek van vegetatieplekken van invloed op de hydrodynamica, omdat meer impuls in de stroomrichting wordt geabsorbeerd door de vegetatie via wrijvingskrachten (lagere sediment transportcapaciteit). Het algehele sedimentatiepatroon en de geulontwikkeling ontstaan door de combinatie van deze twee effecten. Zo blijkt dat de geulpatronen niet alleen bepaald worden door de geulbreedte en het stroomgebied, maar ook door de hoeveelheid aan vegetatie, de frequentie waarmee vegetatiegegevens worden bijgewerkt, de kolonisatiepatronen en andere ecologische eigenschappen. Deze innovatieve bio-geomorfologische modelmatige benadering maakt het mogelijkheid om de dynamische interactie tussen vegetatiepopulatie en de morfologie verandering te bestuderen.

Ten tweede is het model toegepast op de studie van een kwelderherstelproject in het estuarium van de Nisqually rivier, ten zuiden van Puget Sound, VS. Deze studie richt zich op: 1) sedimentatiepatronen in de kwelder en in de riviergeulen; 2) vegetatiepatronen na verwijdering van de dam. Het effect van het vegetatie leidde tot een vermindering van de sedimentatie in de rivier en een toename van sedimentatie op het wadplatengebied daar waar de vegetatie groeit en zich verspreidt. Daar waar sprake is van effecten van vegetatie in het voorheen ingedijkte gebied is het sedimentatie volume significant groter dan in een dergelijk gebied waar vegetatie ontbreekt, hetgeen op zijn beurt weer helpt de vegetatie te verspreiden. Ook voor het bestaande kweldergebied is het erosie volume, nadat de vegetatie is ontwikkeld, veel minder dan in het geval zonder begroeiing. Het distributie patroon van vegetatie soorten is ook kwalitatief voorspeld. De modelresultaten helpen bij het verklaren en voorspellen van de morfologische verandering met invloed van vegetatie populatiedynamiek in kwelders.

Ten slotte wordt het gebruikt om de grootschalige morfologische dynamiek van de shoreface connected radiale zandbanken, gelegen in het zuiden van de Gele Zee, China te bestuderen. Deze studie richt zich op het ontstaan en de evolutie op lange termijn van dit unieke, grote morfologische fenomeen. De modelstudie helpt bij het vergroten van het inzicht in het grootschalige gedrag van de zandbanken. Uit deze studie blijkt dat: 1) het radiale vloedgolfpatroon niet kan worden gegenereerd door alleen de twee gezamenlijke vloedgolven, d.w.z. de vloedgolf door het schiereiland Shandong gereflecteerd en de progressieve golf uit de Oost-Chinese Zee, waarbij ook de flauw hellend bodem van het continentaal plat van essentieel belang is, en 2) de morfologische resultaten van modellen aantonen dat een radiale getijdegolf niet per se tot een groot zandbank systeem met radiale vorm hoeft te leiden.

In het kort samengevat, is een generiek morfologisch model ontwikkeld, dat beschikt over de flexibiliteit om te worden gekoppeld met verschillende hydrodynamische modellen die gediscretiseerd worden tot zowel gestructureerde als ongestructureerde roosters. De open structuur van het model frame en de adaptieve numerieke algoritmen heeft de flexibiliteit van het model mogelijk gemaakt. Door de integratie met de ecologische processen wordt het model uitgebreid tot een bio-geomorfologische model, die verder laat zien van het model uitbreidbaarheid te worden. Het model is gevalideerd tegenover een breed scala aan testcases en is toegepast op rele praktische problemen. De resultaten van deze studie tonen aan dat door onze toegenomen kennis van kustmorfodynamische processen en ecologische processen en de interactie tussen deze processen, wij beter in staat zijn het kustsystem te analyseren en te voorspellen. Dit generieke geomorfologische model kan gebruikt worden als een multidisciplinair onderzoekhulpmiddel voor zowel morfologen als ecologen.

Contents

List of Figures

xxii

xxiii

xxiv

List of Tables

Chapter 1

Introduction

1.1 Background

Geomorphology (in Greek: the knowledge of earth forms) is defined as the study of landforms, including their origin and evolution, and the processes that form them. Geomorphologists endeavor to understand the history, dynamics, and to predict the future changes of land forms due to natural processes and the effects of anthropical impacts. The disciplines involved include geology, chemistry, physics, geodesy, geography, archeology, hydrology, ecology, biology, meteorology, oceanography, civil and environmental engineering, etc.

The coastal area, which is characterized as a highly dynamical system, is the interface between the oceans and continents. The morphology in coastal areas has fascinated people for centuries, not only for safety reasons, for instance, risk of flood, environmental disasters etc., but also for the wish of understanding, and the ability to predict the impacts of the ever growing human interference with the coastal system because of its environmental and socio-economical functions. The improved understanding about the coastal morphological system shows that the coastal behavior is governed by the complex interactions of a number of physical processes and the morphological change which they cause.

In the past century, much research on coastal morphology has been undertaken, not only at long-term (decades to centuries) and large spatial scale (a few kilometers to hundreds of kilometers), such as evolution of estuaries, but also at small scale (millisecond and millimeter), such as entrainment of individual sediment particles. Not only the physical processes, but also biological processes including macro benthic effects and vegetation dynamic effects, are receiving increased attention. Briefly, the research approaches could be classified into: in-site monitoring and field measurement, scaled physical modeling and numerical modeling. Specifically, the dramatic development of the computational power and better software, have led to a variety of mathematical and numerical models. They are claimed to give quantitative descriptions of a range of coastal morphological phenomena. However, due to the complexity of the natural system, there remain significant knowledge gaps concerning not only the constituent processes and their dynamic interactions with the topography, but also the utilization of the models in realistic situations. See De Vriend et al. (1993); Nicholson et al. (1997); Sutherland et al. (2004); Stive and Wang (2003) and Lesser (2009) for reviews of such models.

In the context of coastal geomorphological modeling, an important step being put forward is the introduction of scale separation (De Vriend et al., 1993). Proper modeling practice relies on some level of process aggregation (Lesser, 2009). The following two approaches are commonly applied, namely, the "top-down" approach and "bottom-up approach". The "top-down" approach is based on the hypothesis that a modeling system may not necessarily have to be based directly on complicated, small-scale processes modeling on interaction at corresponding scales of interest (Murray et al., 2008). The "behavior-oriented" modeling (Stive & de Vriend, 1995) is an example of this type, where the general trends and large scale morphological evolutions are modeled while the smaller scale of physical processes and corresponding changes are left out of considerations. On the other hand, "bottom-up" type models attempt to study the important interactions at small scale and combine them to larger-scale, longer-term phenomena. The modeling approach applied in the thesis, process-based type model, belongs to this type. By identifying proper scales of hydrodynamics, sediment transport, morphodynamics, ecological, e.g. vegetation growth processes, a coupled model system in the discipline of bio-geomorphology becomes possible (Ye et al., 2009). To couple the hydrodynamics and morphological processes at various scales, a morphological factor is introduced to accelerate the morphological change. For long term morphological problems, the morphological factor could be time-varying with the wave statistic parameters (Roelvink, 2006; Lesser, 2009; Van De Wegen, 2010; Van De Wegen et al., 2008).

A range of numerical models and related techniques have been under intensive development for the last two decades, such as, coast line models, coastal profile models and coastal area models. Coast line models describe the large-scale behavior (alongshore) after integrating over the small scales (vertical, cross-shore), with the assumption that the coastal profile is maintained constant, but is shifted onshore or offshore with the erosion or accretion of the coastline. The movement of the coastline is caused by gradients in the littoral drift. Refer to Pelnard (1956); Horikawa (1988); Komar (1998); Deigaard and Fredsøe (2002) for one line model, and Bakker et al. (1970) for two-line model and multiline model. Coastal profile models, reviewed by Roelvink and Brøker (1993), ignore the longshore variation, but include the vertical dimension and concentrate on the medium-scale cross shore evolution. The beach states are described either by Deans dimensionless parameters (Wright & Short, 1984), or by Bruuns equilibrium profile (Dean, 1977), or by an empirical profile (Bakker et al., 1970; Kriebel & Dean, 1985), or more sophisticated, by the type of process-based "deterministic" profile models (Roelvink & Brøker, 1993; Schoonees & Theron, 1995). Even though these two types of models had the potential to study the evolution of coastline and beach profiles under specific conditions, the limited number of processes and strong assumptions, such as the assumption of no longshore variation, hamper wide applications. Thus coastal area models are developed. They are 2D or 3D horizontal models consisting of, and linking a set of submodels of the wave field, the tide-, wind-, wave-, and density-driven current field, the sediment transport fluxes and the bed evolution (De Vriend et al., 1993). However, at present stage, fully 3D models describing the currents on a three-dimensional grid are still in a early stage of development, and require excessive computation efforts (Van Rijn et al., 2003). Fully 3D morphological model are far away from practical engineering applications. Thus, two-dimensional depth averaged (2DH) and quasi-3D schemes are focused and put into the most efforts (Coeffe & Pechon, 1982; De Vriend, 1987; de Vriend et al., 1993). Based on the coupling way of the constituent physical processes, De Vriend et al. (1993) categorize this type of models into: initial sedimentation/erosion models (ISE), medium-term morphodynamic models (MTM) and long-term morphological models (LTM). 2DH and quasi-3D coastal area models are successfully applied, such as Cayocca (2001), Wang et al. (1995) etc. The beach nourishment design and evaluation for the Dutch coast is one good example of engineering application of such kind of model (Grunnet et al., 2004; van Duin et al., 2004; Li et al., 2006; Van Leeuwen et al., 2004).

Even though big achievements have been obtained on the topic of coastal morphology modeling, there still remain gaps between the model capabilities and extending engineering demands. For instance, morphological modeling of Texel inlet at Dutch Wadden sea (Elias, 2006), and also modeling of Willapa Bay (Lesser, 2009), showed the significance of including the fully 3D processes, such as channel curvature, secondary flow, residual flow and sediment transport.

In addition to the model capability gap, there are also generality gaps. From model developers' point of view, most existing morphology models are built in as modules within hydrodynamic software packages. As a result, these morphological modules inherit the advantages from the hydrodynamic models, such as, grid discretization, numerical integration methods and coupling schemes etc.. However, limitations are also inherited, e.g. difficulty of maintenance, poor extendibility and compatibility, which restrict the applicability of the morphological modeling. A generic, portable morphological model introduced by this thesis would be a beneficial contribution. Moreover, for geomorphological modeling, there is increasing interest in the influence of plants, animals, and microorganisms on the morphology evolution, which is referred to bio-geomorphology. A bio-geomorphological model which integrates hydrodynamic processes, morphological processes, ecological processes, water quality processes and their interactions at proper scales, is thus requested. Therefore a generic structure for an integrated bio-morphological model is necessary.

1.2 Objectives and methodology

The main objective of this research is to develop a generic morphological modeling approach, and to improve our ability of analysis and prediction of the coastal system based on the increasing understanding of coastal processes and the interactions between them at different temporal and spatial scales.

More specifically, the objectives of this research are:

1. To develop a generic morphological model and to couple it with hydrodynamics models using both structured and unstructured grid.

2. To validate the generic morphological model with morphological process against theories, laboratory data, field data, and other numerical model results.

3. To apply the generic morphological model to study typical morphological features at a wide range of scale.

4. To explore the possibilities to extend the generic morphological model towards a biogeomorphological model and apply it to study coastal morphology development with dynamic effect of vegetation.

1.2.1 Research questions

Four corresponding research questions related to the objectives are imposed.

- From theoretical point of view

1) How could the generic model be coupled with various hydrodynamic models, independent of structured/unstructured grids and the numerical integration methods?

2) How would the generic model represent and couple the dominant coastal sediment transport processes and morphodynamic processes, and possibly, ecological processes?

- From application point of view

3) Could this generic morphological model reproduce typical morphological phenomena at large scales, such as generation, evolution, migration of sand ridge system, and at small scales, for instance, trench migration?

4) Could this generic morphological model reproduce small scale ecological effects on morphological changes in high dynamic environments, such as salt marshes?

1.2.2   Approach

Based on the objectives to be achieved, the research approaches include software development and process-based modeling.

The model development is carried out through the open processes library, an open framework of Delft3D-WAQ. Intentionally, the Open Processes Library is designed as an extensible collection of subroutines used by the water quality module in Delft3D and SOBEK. At present stage, the open process library consists of hundreds of water quality, chemical and ecological related substances and processes. It is an open platform for users to include substance collection with processes by their own interests into the present Delft3D system.

From the developer point of view, the open processes library is a tool to extend the library of substances, and the physical, chemical and ecological processes involved. Once the substances and the related processes are ready, the modeller can define the substances and related processes. After the configuration, the Delft3D-WAQ preprocessor and processor will read the substances, resolve the related processes and maintain the mass balances. In this study, sediment (sand and mud) and vegetation species are added into the substance library, while the corresponding sediment transport processes, vegetation population dynamic processes and interactions between them are added into the process library.

This study starts from model development by decoupling the sediment transport formula and morphological module from Delft3D-FLOW. The sediment transport and morphodynamic processes are implemented through open processes library (Figure 1.1). Afterwards, the model is validated by various test cases with existing analytical, empirical solutions and flume experiments. The hydrodynamics modules in Figure 1.1 is aimed to be 1D/2D flow model Sobek, 2D/3D structured-grid flow model Delft3D-FLOW, and unstructured-grid flow model TELEMAC, or D-Flow FM. Modules and processes included in the red box are newly developed or extended components of the generic coastal morphological model, i.e., the sediment transport module (extended from Delft3D-WAQ), bed state module, bed level update module, and a vegetation dynamic module (the later three are newly developed). The improved modules are added up and verified subsequently. The open structure assures flexibility of the model to be coupled with wave module, aeolian transport module at later stage (see Figure 1.1). The dataflow in the model, including water depth, bed/bank/boundary roughness are described in Section 3.6.

Four modules are developed or extended as the components of the system:
- Bed state module

  Bed state module includes prediction of bed roughness, prediction of suspended sediment size (Van Rijn et al., 2004; Van Rijn, 2007a), graded sediment effects (Van Rijn, 2007c), wave-induced effects, such as, orbital velocity and streaming near bed, bed shear stress, sand transport (Van Rijn, 2007b), Shields criterion for fine sand (Van Rijn et al., 2004), bed book-keeping system (refer to Section 3.2).

5

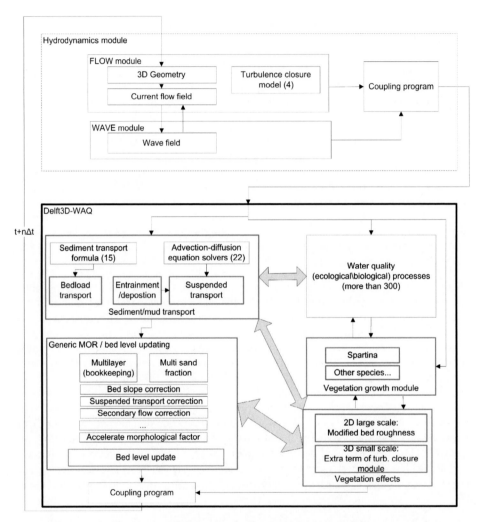

Figure 1.1: Proposed model structure. The content in the red square are to be created/improved in this study.

- Vegetation population dynamics module

  Vegetation population dynamics module includes ecological processes, such as, growth/decay process, spatial spreading and lateral extension, competition/interaction among species, seed dispersion processes.

- Sediment transport module

  Sediment transport module includes 15 sediment transport formula for bedload transport. The suspended sediment transport equation is solved using the numerical schemes in the Delft3D-WAQ, based on finite volume schematization, which is capable for unstructured grid.

  The following are extended for the sediment transport module (refer to Section 3.3):
  1). Density effect of suspended matter;
  2). Sediment exchange between the bed and suspended matter;
  3). Vertical diffusion effects;
  4). Suspended sediment correction;
  5). Bed slope effects for the bedload transport;
  6). Morphological acceleration factor.

- Geomorphological bed updating module

  To develop a generic morphodynamic model, two key things are to be developed, i.e., a generic algorithm for the bed updating using unstructured grid (refer to Section 3.4), and a general velocity integration algorithm using for both structured and unstructured grid (refer to Section 3.5). Furthermore, some efforts is bound to the stability and accuracy for numerical schemes (Callaghan et al., 2006) and effects of different updating algorithms (Roelvink, 2006). Concerns on the morphological factors and mass conservation are discussed further when ecological processes are involved.

The model is applied to study the Jiangsu coastal zone in the southern Yellow Sea, China between the Yellow River and the Yangtze River. This area is an important, dynamic ecological area with unique morphological characteristics in the form of large, shoreface connected radial shape sand ridges. Although the intention was to test the full bio-geomorphological system on this study area, however, since this case is already very complex from a purely morphological point of view, we decided to carry out the morphodynamic analysis with the fast and robust present Delft3D online MOR model, as a necessary first step towards fully coupled bio-geomorphological simulations. The Eastern China Sea model is calibrated and verified based on the field measurements. Discussion on the generation and long-term evolution are carried out by analysis on a schematized continental shelf model (refer to Section 6.4).

The model is also applied to study tidal channels development in salt marshes (refer to Section 6.2). The formation and development of the tidal channel networks are determined by the interactions of hydrodynamic processes and morphodynamic processes. The presence of vegetation adds to the complexity. Following the study of Temmerman et al. (2007), this study offers a generic modeling approach and proved to be a multidiscipline research tool for the future.

The results of this research contribute:
1) to a generic geomorphological model independent of types of discretized grid and numerical methods of the flow modules;
2) to increase the understanding of the tidal sand bank behavior at large scale by data analysis and processes based numerical modeling;
3) to an innovative bio-geomorphological modeling approach by considering the interaction of morphodynamic processes and ecological processes simultaneously.

1.3   Terminology

To minimize the confusion of related terms among the civil engineering community, some definitions of terms in the geomorphological discipline are listed hereby.

- **Geomorphological model**

  In this research, with a geomorphological model, we mean a numerical geomorphological model.

  The numerical geomorphological model uses a set of partial differential equations and boundary conditions, delineating (water, transportable material, sediment in this research) mass balance, momentum balance and energy balance of the physical system, to incorporate a large number of physical processes. Though in a few cases, the set of equations can be solved analytically after simplifications with some strong assumptions, more often the equations cannot be solved directly. Some numerical techniques are employed to solve the set of equations. In this research, a finite volume analogue of finite element schematization is applied (See Section 2.3).

  In coastal zone, the system we are focused on is characterized to be: 1) with free surface; 2) gravity is the most important driven force, where friction from the boundary and internal viscosity dissipate most part of energy; 3) the vertical velocity and vertical acceleration could be ignored. The 3D equations could be integrated vertically to become the so-called shallow water equations (SWE) (Stoker, 1957; Vreugdenhil, 1994).

- **Generic geomorphological model**

  Generic geomorphological model, in this research, is particularly defined in two aspects. One is that the geomorphological model can be coupled with different hydrodynamic models, such as Delft3D-FLOW based on curvilinear structured grid, Telemac based on triangular unstructured grid, Sobek based on a river network schematization, etc.

On the other hand, the generic model is supposed to be robust under different circumstances like the case study areas. Two application case study areas are studied. The radiating-shape sand banks group in the East China Sea is characterized with complicated flow patterns and complex dynamics in a large time and spatial scale as an interesting test case (refer to Section 6.4). The morphodynamic changes in salt marsh in relatively small scale are of ecological interest. The channels on the tidal salt marshes serve as the main paths for flow, sediment, nutrients, etc. The formation of the channel networks is determined by the interactions of tide-induced pressure gradients and residual currents, wind-driven waves, swell, sand/mud transport, sediment sorting in horizontal and vertical direction. The presence of vegetation adds to the complexity. Existence of vegetation could not only slow down the flow and cause sedimentation, but also increase the local turbulence by flow around the stems and leaves, and cause erosion. The root systems introduce strong resistance against erosive force to some depth under the water-sediment interface, but might increase the possibilities of erosion in soil patches under extremely high near-bed shear stress during storms. In Section 6.2, the extended generic bio-geomorphological model is applied to study the morphological evolution and corresponding vegetation spatial pattern in a specific salt marsh.

- **Coupled model / online simulation**

  In the natural system, all the processes are coupled and interacting simultaneously. For instance, the tide driven current, density driven current, wave driven current, etc. affect the dynamic system simultaneously, which is hardly to be simulated in one unique system. In Delft system, the coupled model or a so-called 'online' simulation is used.

- **Delft3D online MOR version**

  Delft3D software system is the fully integrated modeling tool for a multi-disciplinary approach and 3D computations for coastal zone, river, lake and estuarine areas from Deltares (previous WL|Delft Hydraulics). The Delft3D system is composed of several modules, including Delft3D-FLOW for 2D/3D hydrodynamics, transport, WAVE (SWAN) for short wave propagation, SED for cohesive and no-cohesive sediment transport, and MOR for morphodynamic simulations. The Delft3D online MOR is a latest version, for the time being, of Delft software system which allows the online coupling of hydrodynamics, transport, wave, sediment transport and consequent bed level change in each time step (refer to Section 2.2.2 and Figure 2.2). This system is using structured grid and the result of this system is used to verify the new developed generic morphological model system in the later chapters.

- **Structured grid / Unstructured grid (Wenneker, 2003)**

  In *structured grids* the grid cells are quadrilaterals (for 2D cases) or hexahedrons (for 3D cases), and at each interior node four (2D) or six (3D) cells meet, for instance, Cartesian grids, in which the grid cells are rectangles (2D) or blocks (3D), and curvilinear boundary-fitted grids, in which the boundaries of the grid conform to the domain boundaries. A *single-block grid* can be mapped onto a rectangle (2D) or block (3D), which implies that each cell can be uniquely identified by a set of integers (i,j) in 2D or (i,j,k) in 3D. A *multiblock grid* consists of a number of matching single-block grids. Usually there are unstructured connectivities between the blocks.

Unstructured grids subdivide the domain into simple, usually triangular (in 2D) or prismatic / tetrahedral (in 3D), elements with no implied connectivity.

In most practical applications, all grid cells have the same geometric form (triangle or prisms, for example). In a *hybrid grid*, part of the grid consists of quadrilateral (2D) or hexahedral (3D) elements and part of the grid consists of triangular (2D) or prismatic/tetrahedral (3D) elements. Other element shapes are theoretically possible, but hardly used because they do not offer additional advantages while they complicate the underlying algorithms. The reason to use hybrid grids is to mix the advantages of structured (accurate results) and unstructured (flexible grid generation) grids.

Unstructured grids are sometimes miscalled as finite element grids. The finite element method (refer to the above item) is one of the popular numerical methods for solving (partial) differential equations but not a grid generation method. In addition, finite element methods can be applied on structured grids.

- **Tidal sand bank/sand ridges**

  In many shelf seas with strong tidal currents, the offshore seabed exhibits a wide variety of rhythmic bottom features of different length scales, such as sand banks, tidal ridges, sand waves, mega ripples and sand ripples Hulscher et al. (1993). In this research, our attention is restricted to tidal sand banks, which have a typical wave-lengths of 100 times the undisturbed water depth (i.e. kilometers) and crest orientation slightly cyclonic (i.e. counterclockwise on the Northern Hemisphere) up to parallel with respect to the main tidal motion.

  In North Sea, they hardly move and their crests are orientated slightly anticlockwise (angles between 5° and 30°) with respect to the principal tidal current direction. Sometimes, they are superposed by sand waves, which is have typical wave-lengths of ten times the local water depth, their crests are perpendicular to the direction of the principal tidal currents and they migrate several meters per year.

  In East China Sea, they extend $200km$ from north to south, $90km$ from east to west, $0m$-$30m$ depth under theoretical depth datum, and consist of 10 elongated sand banks stretching out from the root part, towards north, northeast, east, southeast, in a typical radiating manner, which is termed "radiating shape" sand banks group. There is still no widely accepted conclusion on whether the sand banks group is stable or not.

  Their formation of tidal sand banks might be explained as a morphodynamic instability of a sandy bed subject mainly to horizontal tidal motion, see Huthnance (1982) and Zimmerman (1981), extended by de de Vriend (1990), Hulscher et al. (1993) and Hulscher (1996).

1.4   Outline of this book

In this research, an important part is related to implementation of the physical and ecolgical processes in the coastal sediment transport and morphological change. The structure of the thesis reflects the objectives outlined above.

In Chapter 1, an overview of the geomorphology context is given. Some terms specified in this research are defined thereby. An outline of the scope of the research is also given. The main objective of this research is presented and specific research questions are followed.

In Chapter 2, a literature review is given. The review focuses on 3 aspects: morphological modeling, research on shoreface connected sand ridges and studies on vegetation effects on morphodynamic change in salt marshes.

In Chapter 3, development and implementation of the generic geomorphological model is described. The model is generic in the following sense: 1) Flexibility. The model could be online-coupled with different kinds of flow model (1D/2D/3D, FDM/FEM, tri/cur grid); 2) Extendibility. The model consists of sediment transport module, morphological module, bed-state description module, water quality module, vegetation growth module, etc.; 3) Open developing environment (user-tailored processes definition); 4) Wide-usability. The model inherits from Delft3D-WAQ, thus mass conservation is fulfilled by default, and hundreds of water quality processes are ready to use. Three modules involved in this research consist of a bed state description model, a sediment transportation model and a morphological bed updating model are to use the finite volume method for unstructured grids.

In Chapter 4, the function of the morphological model is validated against 1) analytical solutions and results from other models, such as: hump migration problem, equilibrium bed slope and sediment concentration profile and 2) flume experiments and results from other models, such as trench migration. Morphological processes, such as suspended sediment transport, bed load transport, bedslope correction and suspended sediment concentration profile, have been validated.

In Chapter 5, development, implementation and validation of the extended part of the generic morphological model towards the bio-geomorphological model are presented.

In Chapter 6, three cases in the coastal areas under complex dynamics are chosen to verify the capability of the generic bio-geomorphological model. The first one is on short term morphodynamics of salt marsh with vegetation effects in relatively small scale, and the second is on middle term morphological evolution with vegetation effects in a salt marsh restoration project in US. The third one is focused on the large scale morphology change at Jiangsu coastal zone in the southern Yellow Sea, China between the Yellow River and the Yangtze River. This area is an important, dynamic ecological area with unique morphological characteristics in the form of large, shoreface connected radial shape sand ridges.

11

In Chapter 7, the key results are summarized and several research points are identified for future studies. The outlines of this thesis are visualized in Figure 1.2.

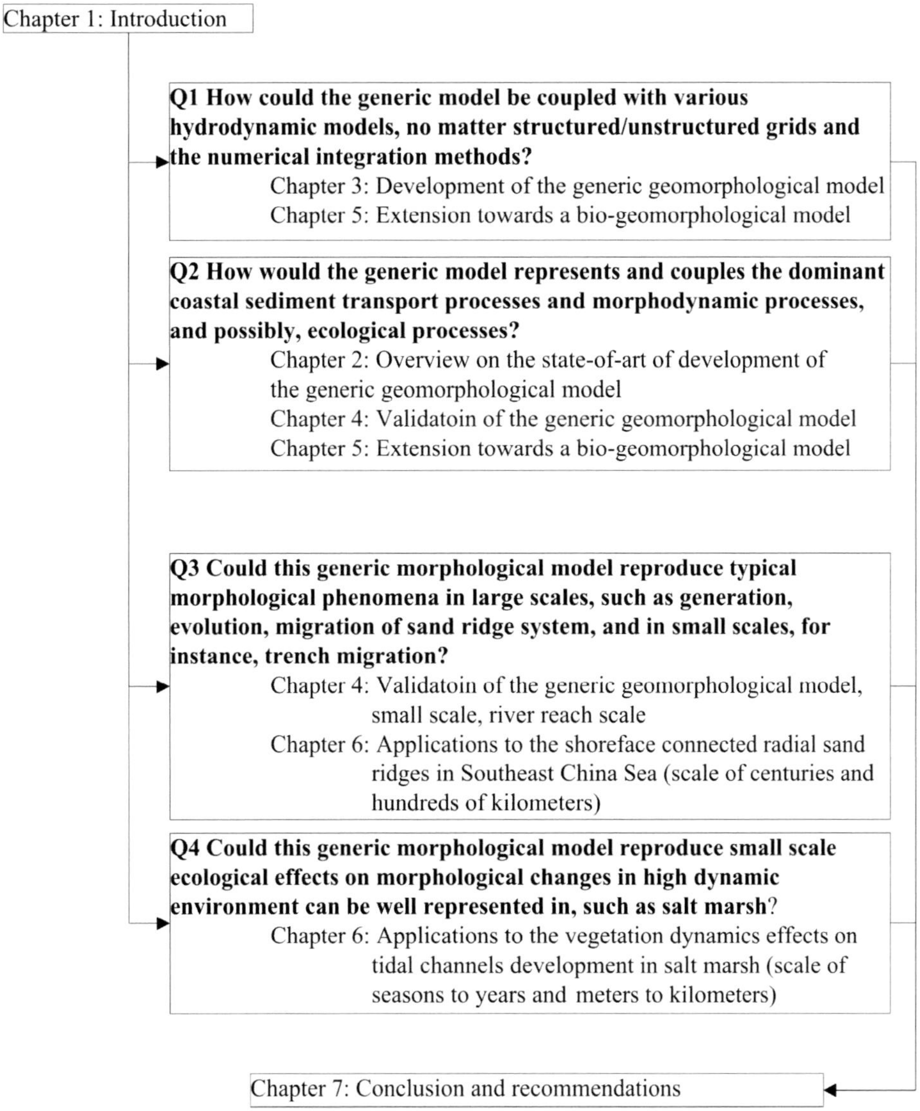

Chapter 1: Introduction

Q1 How could the generic model be coupled with various hydrodynamic models, no matter structured/unstructured grids and the numerical integration methods?
> Chapter 3: Development of the generic geomorphological model
> Chapter 5: Extension towards a bio-geomorphological model

Q2 How would the generic model represents and couples the dominant coastal sediment transport processes and morphodynamic processes, and possibly, ecological processes?
> Chapter 2: Overview on the state-of-art of development of the generic geomorphological model
> Chapter 4: Validatoin of the generic geomorphological model
> Chapter 5: Extension towards a bio-geomorphological model

Q3 Could this generic morphological model reproduce typical morphological phenomena in large scales, such as generation, evolution, migration of sand ridge system, and in small scales, for instance, trench migration?
> Chapter 4: Validatoin of the generic geomorphological model, small scale, river reach scale
> Chapter 6: Applications to the shoreface connected radial sand ridges in Southeast China Sea (scale of centuries and hundreds of kilometers)

Q4 Could this generic morphological model reproduce small scale ecological effects on morphological changes in high dynamic environment can be well represented in, such as salt marsh?
> Chapter 6: Applications to the vegetation dynamics effects on tidal channels development in salt marsh (scale of seasons to years and meters to kilometers)

Chapter 7: Conclusion and recommendations

Figure 1.2: Outlines of this thesis

Chapter 2

Geomorphological Modeling: An overview

2.1 Introduction

Geomorphological modeling has proven to be an efficient approach to study coastal geomorphology. Numerous research groups have put in significant efforts to develop comprehensive morphology models. This chapter can only provide a very brief overview of scales, processes and numerics. The chapter starts in Section 2.2 with a review of scales of geomorphodynamic systems and corresponding models. Subsequently, we will focus on the process formulations. There are several complex morphological systems, such as, TELEMAC ("Morphodynamic modeling using the Telemac finite-element system", 2011), CSTM/ROMS ("Development of a three-dimensional, regional, coupled wave, current, and sediment-transport model", 2008), XBeach ("Modelling storm impacts on beaches, dunes and barrier islands", 2009), MIKE system from DHI and Delft3D system from Deltares. However, all these models have been developed specifically for either structured grids or triangular meshes, whereas integrated modeling studies increasingly demand a more flexible mixture of 1D, 2D and 3D modeling capabilities on a combination of triangles and quadrilaterals. The geomorphology modules that will be developed in Chapter 3 are intended to provide that flexibility. With respect to the process formulations, the new modules will be intensively compared against the existing Delft3D online MOR system. Therefore, a brief overview of the geomorphological processes and implementation on structured grids in the standard Delft3D online MOR version is given in Section 2.4.1. Finally, this chapter concludes in Section 2.4 with an overview of the most widely used models and their numerical methods. The newly developed generic morphological model described in the next chapter is based on a finite volume core suitable for arbitrary geometries of grid cells. Compatibility of finite element and finite volume methods has been shown by Postma and Hervouet (2007) which indicates that the chosen approach should be compatible with a large number of hydrodynamic models, such as TELEMAC.

2.2   Morphodynamic system

Coastal process-based morphodynamic models are used to analyze and predict the changes of the morphodynamic system at different spatial and temporal scales, from dune erosion (hours to days time scale and 10 meters spatial scale), to effects of groynes and breakwaters, nearshore bar generation (weeks to months and 100 meters to 10 kilometers), tidal inlets, river mouths, estuaries and bays (years to decades and large spatial scale). Applications of the generic geomorphological model described in Chapter 6 include long term morphodynamics of sand banks at large scale and short term morphodynamics of salt marsh at relatively small scale. In this section, scales of the morphodynamic system are discussed and then corresponding types of model are reviewed.

2.2.1   Scales for the morphodynamic system

It is always necessary to distinguish different scales in the system before using models, because the representations of dominant processes in different models are always different.

Hydrodynamics processes considered in this study mainly include tidal current, short waves, and turbulence. Each of these processes has its own characteristic range of scales. Tidal currents are a dominant force in the offshore area. At the largest spatial scale, the tidal current is regarded as a type of long wave, propagating along the continent shelf, with length of hundreds of kilometers and period of around 12 hours (semi-diurnal tides) or 24 hours (diurnal tides). At longer time scales equilibrium theory predicts the variation in tidal range of the spring-neap cycle with a period of 14.7 days. Furthermore, there is a seasonal (yearly) cycle, governed by the rotation of the Earth around the Sun with a period of 365.25 days. Even longer period cycles are the 18.61 year nodal cycle due to the revolution of the moon's nodes and the 1600 to 1800 year cycle due to astronomical alignments (Marchuk & Kagan, 1984). At smaller spatial scales, the tidal wave can break due to favorable bathymetry which results in a tidal bore (Dronkers,

2005). Typical maximum tidal current speeds are approx. 1 m/s (in tidal inlets) and up to approx. 3 m/s (in tidal barriers).

Short waves play a significant role too in coastal dynamics at a range of scales. There are several categories of waves, such as, capillary waves (with a wave height less than 2 cm), wind-generated short crested gravity waves (with periods of 5 to 10 seconds and wave length of tens of meters), swell waves (with periods of 10 to 16 seconds and wave length of 150 m), storm waves (caused by a change in pressure, with wave height of up to 26m (Haver, 2004)), surf beat (a collective term for low frequency gravity waves) or infragravity waves (caused by wave groups breaking in the surf zone, with periods from 20 to 200 s), harbor resonance waves (periods in the range of 2 to 10 minutes), and tsunami waves (by sub-sea earthquake, with wavelength in excess of 100 km and period in the order of one hour).

In turbulent flow regime, which often includes rapid variation of pressure and velocity in space and time, unsteady vortices emerge at various scales, interact with each other, and form eddies in a hierarchy of length scales (Pope, 2000; Tennekes et al., 1972). The smallest (mean) length scale at which energy dissipation happens is the Kolmogorov length, defined as $\eta = (\nu^3/\varepsilon)^{\frac{1}{4}}$, where ν is the viscosity, ε denotes the rate of energy dissipation. With a velocity scale defined as: $u' = \sqrt{k}$ and the mean kinetic energy in the turbulence k defined as: $\frac{1}{2}\langle u_i' u_i' \rangle$, the corresponding time scale of turbulence is then $t = \eta/u' = (\nu/\varepsilon)^{\frac{1}{2}}$. Between the smallest scale and the typical flow scale L, there is a cascade of scales r to pass on the energy, i.e. $\eta < r < L$. For uniform open channel flow, the most important length scale of turbulence is the boundary layer thickness at the bed. The characteristic time scale of measured turbulence in small subtropical estuaries with semidiurnal tides is around 0.06 and 2 s (Trevethan et al., 2008). For modeling flow through the vegetation, Uittenbogaard (2003) identified the intrinsic time scale as:

$$t_{int} = \frac{k}{\varepsilon} \tag{2.1}$$

and geometrical-imposed time scale as:

$$t_{geo} = \left(\frac{L_p^2}{c_\mu^2 T} \right)^{\frac{1}{3}} \tag{2.2}$$

where the L_p = length of vegetation, T = power as (production) source term of k equation.

Sediment transport rate generally adapts quickly to local flow conditions and sediment properties. It tends to approach equilibrium ($c = c_e$) in steady flow. Therefore, the scales of sediment transport mainly depend on the flow dynamics. However, for detailed modeling it might be necessary to include the adaptation time of the suspended sediment to reach the equilibrium status, which is denoted as (Gerritsen et al., 1998, Galappatti, 1983, Wang et al., 1992):

$$T_s = \frac{h}{\omega_s}\left(\frac{1}{1+\beta}\frac{\gamma_1}{\gamma_0}\right) \tag{2.3}$$

where: β is the bed level height ratio z_a/h; z_a is defined as bed load transport layer thickness; h is the overall water depth; the appropriate profile coefficient γ_1/γ_0 is a polynomial function of ω_s/u_* (Wang, 1992).

Morphology can be regarded as the response of the bed to hydrodynamic and sediment transport processes. Cowell and Thom (1997) categorized the time scales at which coastal morphological processes operate into four classes:

1. Instantaneous time scales: from a few seconds to days or weeks. Morphological evolution occurs during a single cycle of the forces that drive the morphological change (waves, tides). For example, the destruction of wave ripples under a group of high waves, onshore migration of intertidal bars over a single tidal cycle.

2. Event time scales: from a few days to years. Coastal evolution reacts to processes operating across time spans ranging from that of an individual event, through to seasonal variation in driving forces. For example, the scarping of coastal dunes in response to a major storm, the seasonal closure of an estuary by a sand bar.

3. Engineering time scales: from a few months to decades. Many fluctuations in the driving forces result in coastal evolution that takes place over engineering time scales; this is the time scale that coastal engineers are most concerned with and ranges months to centuries. For example, the migration of tidal inlets, the development of a foredune ridge.

4. Geological time scales: time scales operate over decades and longer. Whereas at the previous three time scales morphological change results from fluctuations in the driving forces, on geomorphological time coastal evolution occurs more in response to the mean trends in the forces (sea level rise, climate changes), such as, the infilling of a tidal basin or estuary, onshore of a barrier system and the switching of delta lobes.

It may be concluded that larger scale phenomena are always accompanied by long term processes (Figure 2.1)(De Vriend, 1991). For example, whether the Dutch Wadden Sea over recent 400 years has reached its equilibrium could be questioned, especially after the extensive construction of coastal defense works on the southern shore of the inlet in 1750 AD and the damming of the Zuiderzee, the major part of the back-barrier basin in 1932 AD interrupted the natural morphological evolution of the ebb-tidal delta (Elias & Van Der Spek, 2006; Elias, 2006). Due to the damming of the Zuiderzee, the tidal channels might require the adaptation time as long as a century before the channels and hence the tidal inlet system reach a new morphological equilibrium (Kragtwijk et al., 2004). Also physical processes at smaller scale are identified in corresponding spatial scale. For instance, morphology change in the Frisian Inlet, one of the inlets of the Dutch Wadden Sea, is dominated by the combined stirring of sediment by tides and subsequent transport by residual currents and tidal asymmetry. The stirring of sediment by tides dominates on the seaward side of the strait whereas tidal asymmetry is the dominant

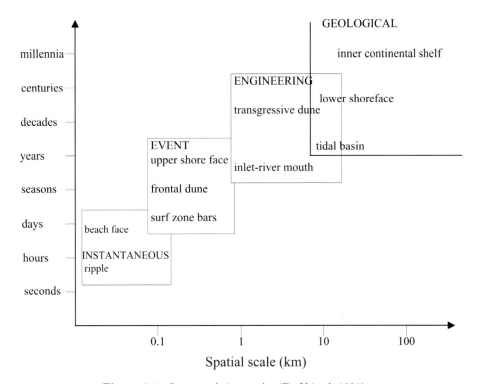

Figure 2.1: Space and time scales (De Vriend, 1991)

factor in the basin. Sediment deposition occurs on both the seaward and landward sides of the barrier islands. The long-term development of that scale of the inlet reaches equilibrium after a hundred years (Van Leeuwen et al., 2003). There is another example for meso-scale morphology change. A meso-tidal sand-gravel inlet at the mouth of the Deben estuary, southeast England was studied. Burningham and French (2006) analyzed the behavior of the inlet and ebb-tidal delta over the last 200 years using a so-called "ebb delta breaching" model developed by FitzGerald (1988). And it was found that the estuary inlet is historically dynamic, with ebb-tidal shoals exhibiting broadly cyclic behavior on a 10 to 30 years timescale. The most recent cycle (1981-2003) indicates the sediment bypassing mechanism operating in the gravel-dominated or gravel-rich inlets inlet is comparable to that in the sand-dominated systems, although the scales and rates of change exhibited are notably smaller and the bypassing cycle is larger due to its coarse-grained sedimentology and the lower efficiency of sediment transporting processes, which implies that the morphological equilibrium may not exist in this specific system.

Some other studies have been carried out at mid/short term and meso/micro scale, for instance, Cañizares et al. (2003) employed input filtering technique, and used a representative tide of the spring/neap cycle and eight representative wave conditions for the different components of the morphological model: hydrodynamics, waves, and sediment transport, which were validated and integrated into the morphological model for the changes of the main morphological features at Shinnecock Inlet in US; Vila-Concejo et al. (2003) developed a conceptual evolutionary model for the natural evolution of artificially opened inlets and analyzed a 2-year cyclic morphological change, including inlet channel evolution (in terms of width, depth and cross-sectional area) and tidal delta formation at Ancão Inlet, located in the Ria Formosa barrier island system of southern Portugal, since the artificial opening in June 1997. Especially for storm-induced morphological change, many other studies have been carried out at short term, for instance, "Modelling storm impacts on beaches, dunes and barrier islands" (2009), Van Rijn et al. (2003), Aagaard et al. (2005), Meulé et al. (2001).

These examples show that the response of the geomorphological systems is corresponding to the spatial and temporal scales and the physical processes at those scales. On the other hand, it should always be kept in mind that since the natural system is nonlinear, its behavior may even be inherently unpredictable in long-term and large scale, and if this is the case, the validity of long-term model predictions does not follow straightforwardly from the model's validity at the hydrodynamic time scale. Processes which are negligible at the smaller scale can have significant long-term effects and vice versa(de Vriend et al., 1993). Our knowledge on which short-term processes are important in the long run is still insufficient. Hibma et al. (2003) showed how, due to non-linear interactions, a simple and regular pattern of initially grown perturbations merges to complex larger-scale channel/shoal patterns with two case studies in the Western Scheldt, the Netherlands, and in the Patuxent River estuary, Virginia. Hibma et al. (2004) succeed in bridging meso- and macro scale processes by reproducing meso scale patterns in a macro scale evolution, which shows positive feedback processes leading to self-organization derived from physical principles on smaller scales. Her work can be regarded as the first successful try in aggregation of meso scale estuarine process scales to macro ones. Geomorphological processes act on time scales ranging from microseconds, which are relevant for turbulent velocities, up to millions of years for geological processes. The spatial scales are similarly wide, from millimeters for capillary flows among sediment particles, up to the continental and global scales (Baptist, 2005).

2.2.2   Models used for processes at different scales

Different types of numerical models are applied to simulate different processes at different scales. Various types of morphodynamic models reviewed by De Vriend et al. (1993) are listed as the following:

1.   Coastline models: including one-line model (Pelnard, 1956; Larson et al., 1987; Horikawa, 1988; Komar, 1998; Deigaard & Fredsøe, 2002) and multi-line model (Bakker et al., 1970). Coast line models describe only the large-scale behavior (alongshore) after integrating over the small scales (vertical, cross-shore). The basic assumption is that the coastal profile is maintained constant, but is shifted onshore or offshore with the erosion or accretion of the coastline. The movement of the coastline is caused by gradients in the littoral drift.

Pelnard (1956) and Larson et al. (1987) developed a range of analytical solutions, for beach evolution with and without the influence of coastal structures, such as, arbitrary initial shapes, sand mining, river discharges, groynes and jetties, detached breakwaters, and seawalls, to this coastline evolution equation, which is simplified as a heat conduction equation, assuming the variation in the coastline orientation small enough to make a linearization and taking diffusion coefficient as a constant. Falqués (2003) analyzed the diffusivity coefficient in detail. And Falqués and Calvete (2005) also extended the theory by taking into account the curvature of the coastline features and instability influence of high waves with long periods and high incidence angles. Falqués et al. (2000) used this method in solving the problem coupling between topographic irregularities, which leads to excess gradients in the wave radiation stress and causes a steady circulation, and wave-driven circulation creates a sediment transport pattern that reinforces the bottom disturbance and changes the large-scale bed forms, such as, giant cusps, crescentic pattern, etc. in the surf zone. Klein (2006) analyzed the linear and nonlinear free-behavior of coastal systems with and without shore face nourishments in Egmond, the Netherlands.

This type of model approach is also termed linear or nonlinear stability analysis. In spite of the incomplete understanding of the mechanics of sediment transport, some essential mechanisms operating at large scale and long-term morphodynamics are able to be clarified by this model approach (Seminara, 1998).

UNIBEST-CL+ is a good example as a one-line coast model tool, which implement this theory. GENESIS, LITPACK, and SAND94 from IBW-PAN software are of the same type. These models are compared by studying the shoreline evolution of the coastline at Wladyslawowo harbor in Poland, characterized by distinct alongshore sediment transport (Szmytkiewicz et al., 2000).

2. Coastal profile model: ignoring the longshore variation. Reviewed by Roelvink and Brøker (1993), this type of model includes:

i) Dean's dimensionless parameters (Wright & Short, 1984), to determine the transition of the beach state and to classify beach states, such as, dissipative or reflective etc.

ii) Bruun's profile (Dean, 1977), which is based on the conservation of wave energy assuming that the average beach profile tends to an equilibrium.

iii) Multi-line model in long-shore direction and equilibrium profile models in cross-shore direction, such as, Bakker et al. (1970); Kriebel and Dean (1985). The strength of the type of models lies in the limited computing capacity required and in the possibility to combine these models with simplified long-shore morphological models.

iv) "Deterministic" profile models, explicitly take into account the different processes (Roelvink & Brøker, 1993). The basic principles are to compute the sediment transport distribution over the profile as a function of the cross-shore profile, sediment properties and seaward boundary conditions, such as wave height and period.

Examples of modeling tools in the type include: *Nearshore Profile Model* from HR Wallingford Ltd; *UNIBEST* from Deltares; *LITCROSS* from Danish Hydraulic Institute; *SEDITEL* from Lab. Nationale d'Hydraulique; *WATAN3* from University of Liverpool, and *REPLA* from SOGREAH etc. (Roelvink & Brøker, 1993; Brøker Hedegaard et al., 1992). Schoonees and Theron (1995) analyzed them and compared based on the theoretical basis (mainly sediment transport) and the associated verification data (mainly morphodynamics) for short-term and mid-term applications, and concluded that the models had the potential to study the evolution of a beach profile under specific conditions. However, the limitation of the models hampered the application of coastal cross-shore profile model, such as, some processes, for instance, the swash zone processes, cross-shore sediment transportation and dune erosion, are not described in the models. Some mechanisms, like the velocity field in the area just before and after the break point, cross-shore distribution of longshore sediment transport, the coupling of effects of long scale longshore sediment transport and short scale of cross-shore sediment transport etc., are still understood rather poorly. It is assumed that the evolution of the beach is relatively long enough to achieve the equilibrium profile.

However, even though the profile models have limitations in engineering applications, they are still good tools for large scale analysis.

3. Coastal area model: 2D or 3D horizontal models consisting of, and linking, the same set of submodels of the wave field, the tide-, wind- and wave-driven flow field, the sediment transport fluxes and the bed evolution (De Vriend et al., 1993). Two-dimensional depth averaged (2DH) schemes (Coeffe & Pechon, 1982; De Vriend, 1987) are reviewed by De Vriend et al. (1993). He listed three basic categories of the coupling way of the constituent physical processes (Figure 2.2):

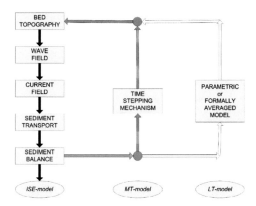

Figure 2.2: Three categories of compound morphological model concepts(De Vriend et al., 1993)

i) Initial sedimentation/erosion (ISE) models
   Only one sweep on the sequence of constituent models was carried out. In fact, the hydrodynamic and sediment transport computation is based on the assumption of invariant bed topography and only the rate of sedimentation or erosion for that topography is computed at every location. (Steady Assumption)

ii) Medium-term morphodynamic (MTM) models
   The new bottom topography is fed back into the hydrodynamic and sediment transport computations; this yields a looped system which describes the dynamic time-evolution of the bed. The timescale of this essentially deterministic "morphodynamic" simulation cannot be substantially larger than the hydrodynamic time scale (duration of a storm, tidal period), even allowing for future improvements to the efficiency of time-stepping techniques.

iii) Long-term morphological (LTM) models
   The constituent equations are not describing the individual physical processes, but integrated processes at a higher level of aggregation. Boundary conditions are also lumped to parameterize the interaction of individual physical processes.

Coastal area models are successfully applied, such as Cayocca (2001),Wang et al. (1995),etc. Other examples, like the beach nourishment design and evaluation in Dutch coast applied such kind of models successfully (Grunnet et al., 2004; van Duin et al., 2004; Li et al., 2006; Van Leeuwen et al., 2004).

4. Quasi-3D model: a combination of area model and analytical description along the vertical plane. De Vriend and Stive (1987) presented a quasi-3D model on the basis of existing concepts of 2DH (horizontal plane) and 2DV (vertical plane) current models. Briand and Kamphuis (1993) listed a quasi 3D numerical model for sediment transport in the surf zone and swash zone, thus the beach profile changes, by combining a 2DH hydrodynamics model and 2DV cross-shore profile model. Katopodi and Ribberink (1992),

and Rakha (1998) showed some comparison on 1DV, 2DH and quasi-3D modeling as well. Drønen and Deigaard (2007) developed a morphological quasi-3D area model combining a two-dimensional depth integrated model for normal and oblique direction wave-driven currents with a model for undertow circulation currents. Their combined model makes a simultaneous simulation of the bar-forming processes associated with the undertow and the horizontal wave-driven circulation currents, which may cause instabilities of the bar and the formation of rip channels. For normal incident waves the quasi-3D model produces a crescentic bar while the depth integrated model predicts almost straight sections of the bar interrupted by rip channels, which implies that the nonlinear morphological system with more processes may lead to different results.

5. Local model: also termed as 1DV model. In some specific issues, 1DV models are employed to solve problem at relative small scale, for instance, Van Ledden (2003)'s work on sand-mud segregation at hot spots in estuaries and tidal basins. DELCON, a one-dimensional (vertical) hydrodynamic model, is used to simulate the settlement of soils, which focus on turbulence, density currents, sediment settling, deposition and consolidation, gas bubble nucleation processes (Winterwerp et al., 1997).

In summary, enormous efforts have been put into the morphological modeling techniques. For specific problems, dominant processes would be identified and different types of numerical models are then applied based on the corresponding scales. In this research, the generic geomorphological model belongs to the type of 2D/3D area model.

Furthermore, in recent years, several new techniques are introduced to the geomorphological modeling practice, which is briefly described in the following.

New techniques for coastal morphodynamics models

Recently, with new technologies developing, more and more new methods are employed for the coastal morphodynamic modelling.

• Data driven modeling
Rózyński (2003),Kingston et al. (2000),Pape et al. (2007) used data driven modelling (such as, artificial neural network) to describe multiple longshore bars and nearshore sandbar behavior.

Their strength thus lies in the ability to make hindcasts and potentially, forecasts in situations where process-based models cannot be applied or will fail. However, the potential for predictions is at the expense of generality and process-knowledge improvement. For instance, it is unlikely that the data-driven models for the temporal evolution of the intertidal momentary coastal line at Egmond will produce any sensible predictions elsewhere (Smit et al., 2007).

• Data assimilation

Data from satellite image, remote sensing, radar data and Argus video data are used for data integration into the models.

Scott and Mason (2007) investigates the use of data assimilation to enhance a simple 2DH decoupled morphodynamic model to predict the waterline based on SAR satellite image which is essentially a contour of the bathymetry at some level within the inter-tidal zone of the Morecambe bay over a 3-year period. The results support that the use of data assimilation successfully compensates for a particular failing of the model, and helps to keep the model bathymetry on track and improves the ability of the model to predict future bathymetry, which suggests that data assimilation should be considered for use in future coastal area morphodynamic models.

Some other new initiatives to implement the data assimilation technique to improve the ability of the coastal morphodynamic models(Smit et al., 2007), such as, Beach Wizard to assimilate dense remotely-sensed field observations of wave dissipation, phase speed, surface currents, water line position, and sand bar morphology into the process-based morphological model Delft3D to provide integrated bathymetric and hydrodynamic nowcasts and short term (order one week) forecasts (van Dongeren et al., 2008). Another example is that the project "Predictability and uncertainty analysis of nearshore sandbar behavior" aims to study the predictability of nearshore sandbar dynamics by applying novel approaches to quantify uncertainties in nearshore sandbar behavior associated with parameter and observational errors, and to integrate 2D bed-evolution process models with remote-sensing data of nearshore bathymetry and to assess short-term variability in the model's parameters (Ruessink, 2005).

2.2.3   Evaluation of coastal morphodynamic models

Evaluating the performance of numerical models of coastal morphology against observations is an essential part to establish their credibility.

The most common quantification approaches used to determine the quality of a morphodynamic model are statistic characteristics, including the following 2 types:
  1. Raster way (include Error Statistics and Threshold Statistics):

  - Accuracy (Sutherland et al., 2004): including the Mean Absolute Error, the Root Mean Square Error (RMSE) and the Mean Square Error.

$$RMSE = [\langle(\Delta h_{model} - \Delta h_{measured})^2\rangle]^{0.5} \qquad (2.4)$$

  - Bias (Sutherland et al., 2004): the difference in the means.

$$Bias = \langle(\Delta h_{model} - \Delta h_{measured})\rangle \qquad (2.5)$$

- Average geometric deviation: defined as the following:

$$Agd = \left(\prod_{i=1}^{n} R_i \right)^{\frac{1}{n}} \tag{2.6}$$

where: $R_i = max(h_{i,model}/h_{i,measured}, h_{i,measured}/h_{i,model})$ as spatial discrepancy.

- Kappa statistic (Cohen, 1960; Tang et al., 2009): A category table for the model simulated variables and observations, i.e., water level, vegetation density, bed level etc., are counted grid by grid. It is a widely used coefficient to quantify the agreement from two sources (in this study, the two sources are measurement results and observation results).

$$\kappa = \frac{p_0 - p_e}{1 - p_e} \tag{2.7}$$

where: p_0 is defined as summation of diagonal values of the evaluation matrix ($=\Sigma_{i=1}^{c} p_{ii}$), p_e is defined as the probability of random agreement ($=\Sigma_{i=1}^{c} p_{i,row} p_{col,i}$), and c is the number of categories. Examples of the matrix will be given in Section 5.5.

- Brier Skill Score (BSS) (Sutherland et al., 2004): to be described below .
- Probability of Detection (PoD): to be described below.
- Critical Success Index (CSI), also called Threat Score (TS): to be described below.

2. Rector way:

- Fractural pattern (Hibma et al., 2004).

A Brier Skill Score (BSS) is suggested as:

$$BSS = 1 - \frac{\langle (Y - X)^2 \rangle}{\langle (B - X)^2 \rangle} \tag{2.8}$$

where:
Y is a prediction;
X is a observation or benchmark run results;
$\langle , \rangle$ denote arithmetic means;
B is a baseline prediction or a status of zero morphology change. Thus the BSS definition

changes to be (Lesser, 2009), or (Van De Wegen, 2010):

$$BSS \;=\; 1 - \frac{\langle(\Delta h_{model} - \Delta h_{measured})^2\rangle}{\langle(\Delta h_{measured})^2\rangle} \qquad (2.9)$$

$$BSS \;=\; 1 - \frac{\langle(\Delta vol_{model} - \Delta vol_{measured})^2\rangle}{\langle(\Delta vol_{measured})^2\rangle} \qquad (2.10)$$

To include the measurement errors in the BSS, Van Rijn et al. (2003) suggested a form of:

$$BSS \;=\; 1 - \frac{\langle(|\Delta vol_{model} - \Delta vol_{measured}| - \delta)^2\rangle}{\langle(\Delta vol_{measured})^2\rangle} \qquad (2.11)$$

where:
δ is the measurement error (m^3).
When $|\Delta vol_{model} - \Delta vol_{measured}| < \delta$, $|\Delta vol_{model} - \Delta vol_{measured}| - \delta = 0$.

Finally the BSS criterion is classified as 2.1:

Table 2.1: BSS criterion (Van Rijn et al. 2003)

	BSS	BSS_{VR}
Excellent	0.5-1.0	0.8-1.0
Good	0.2-0.5	0.6-0.8
Reasonable	0.1-0.2	0.3-0.6
Poor	0.0-0.1	0.0-0.3
Bad	<0.0	<0.0

For spatial pattern, Probability of Detection (PoD) is widely used in image process-ing and meteorology (e.g., http://origin.hpc.ncep.noaa.gov/npvu/confpres/hydromet02 /hydromet02_1.pdf). It is a likelihood, expressed as in percentage, which a model or detecting method correctly represents the measurement or the benchmark results. The general form is:

$$PoD \;=\; \frac{Ac}{Ac + Am} = \frac{Hits}{Obs} \qquad (2.12)$$

where: A_c is the correctly predicted area, and A_m is the miss shooting area.

An alternative form is named Critical Success Index (CSI), also called Threat Score (TS), also widely used in meteorology (e.g., http://www.hpc.ncep.noaa.gov/research/amsver

27

/sld007.htm):

$$TS \; = \; \frac{Ac}{Af + Ao - Ac} = \frac{Hits}{Forecast + Obs - Hits} \tag{2.13}$$

where: A_c is the correctly predicted area, A_f is forecasting area and A_o is the observation area. These three indexes have been used to access the simulated spatial pattern of macrophytes population in Lake Veluwe in Section 5.5.3.

The other ways to evaluate the modeling capacity are the indexes with more physical meaning, including the hypsometry, tidal volume, width/depth ratio, tidal amplitude/channel depth, shoal volume/channel volume, inlet cross-section area/tidal prism volume (Dronkers, 2005). Those indexes have been applied in Chapter 5 and Chapter 6 to evaluate the morphology change.

2.3   Development of geomorphological models

Generally the modern geomorphological area model should consist of several elementary modules, e.g., a bed state description module, a hydrodynamics module, composed by a wave module, and a current module, a sediment transport model, and a bed level updating module etc.(Pan et al., 2007). This newly developed generic geomorphological model consists of 6 modules, i.e., a wave module and a current module, a bed state module, a sediment transport module and a bed level update module, and a vegetation population dynamic module.

For the current and wave module, the flexible flow module D-Flow FM coupled to structured or unstructured SWAN is under development by Deltares (Kernkamp et al., 2011). Since this flow module hasn't been completed yet, we have used the existing modules Delft3D-FLOW and WAVE instead. To some aspects this would also prove that the newly developed morphological modules are generic to be coupled with hydrodynamic modules using structured (curvilinear and/or spherical) grid. A brief overview about the general governing equations of flow, wave, sediment transport and morphology are given in Section 2.3.2. Numerical schemes of D-Flow FM using unstructured grid and unstructured SWAN is overviewed in Section 2.4.2.

The later four modules are newly developed. A brief overview on the state-of-the-art of the bed state models, the sediment transport module and the bed level update module are also given below. Review on the vegetation population dynamic module will be given in Chapter 5.

2.3.1   Bed state module

New high-resolution measurements using acoustic altimeters and side scan sonar could provide more accurate bathymetry data than before. Even the 3D character of bed forms at high temporal and spatial resolution can be captured. For example, data from airborne Lidar system generally could be 3 meters by 3 meters resolution with vertical accuracy of tens of centimeters. High resolution measurement of bed state gives the possibility to setup and validate the bed state module. In this study, some attempts have been carried out to setup a bed state module for

The sea bed form presents the interaction of various forces. The essential forces include wave and current. The local current could be classified by the Reynolds number $Re(\frac{U_\delta A_\delta}{\nu})$ and relative thickness$(\frac{A_\delta}{k_s})$, as laminar flow and turbulence flow, which could be further described as two sub categories, namely the hydraulic smooth regime and hydraulic rough regime (Sleath, 1984; Van Rijn, 1993). Waves lead to oscillatory flow, and the breaking wave transports mass and momentum, which add the complexity (and beauty) of the system consequently. Different flow condition results into various types of bedforms (Van Rijn, 2007a, 2007b, 2007c, 2007d).

Briefly, in steady flow:
1) Lower transport regime with flat bed, ripples, dunes, and bars;
2) Transitional regime with washed-out dunes and sand waves;
3) Upper transport regime, Froude number $Fr > 0.8$ and Shield number $\theta > 1$ with flat mobile bed and anti-dunes; sand waves with $Fr < 0.8$ and $\theta > 1$.

In tidal estuaries, asymmetric mega ripples, and weakly asymmetric or symmetric sand waves are the most frequently observed.

In coastal areas (oscillatory flow with weak current), vortex ripples (Bagnold, 1966; Sleath, 1984), short wave ripples (SWR, when mobile number $50 < \psi < 150$, ψ is defined below) and long wave ripples (LWR, $\psi > 150$) are dominant. The number ψ depends on the particle size d_{50}, wave period T_p, and peak near-bed orbital velocity U_w.

$$\psi = \frac{U^2_{w,peak}}{(s-1)gd_{50}} \tag{2.14}$$

The crestline pattern of the ripples, sandwaves etc. could be straight, sinuous, catenary, linguoid, lunate and cuspate(Sleath, 1984; Van Rijn, 1993).

The seabed could be permeable, if there is air trapped in the pores, bed slope effects, bed material composition, well-grained or not, which will add the complexity as well.

Though much effort has been put into the sea bed description, there are still some mechanisms that remain unresolved. For instance:

1. Relations between large scale bed forms and small scale bed forms

   Thornton et al. (1998) analyzed the small-scale morphology across the surf zone and found variability of bed forms. For example, 10 to 40 cm high lunate and straight-crested mega ripples are often seen on the seaward flanks of bars, in the nearshore trough, and in rip channels, but their origin and spatial variability are not understood.

   Hay and Wilson (1994) observed the temporal evolution of bed forms at fixed location during a storm based on rotary side scan images. Transition between bed form types occurs on time-scales comparable to the time-scales of changes in fluid forcing, but is also linked to bed form scale and forcing history. Under large waves, significant changes in small-scale bed forms can occur within a single wave cycle (Hanes et al., 1998).

   In contrast, large-scale bed forms can exhibit significant hysteresis in their temporal evolution. These complexities in bed form development are hardly included in the existing models for sediment transport or hydrodynamics.

2. Effects of various sediment fractions with spatial variations

   Variation in sediment size may contribute to the high variability of bed forms in the nearshore. Cores through a storm-deposited bar at Duck revealed grain size variations from a few millimeters thick cross-bedded laminae of grains having diameters two to three times the mean grain size, to several centimeters thick horizontal strata of coarse sand and fine gravel. The temporal and spatial variability of grain size is greatest in the swash zone, where sediment varies from fine sand to gravel and cm-long shell fragments over distances of tens of centimeters and over times order of individual swash excursions.

3. Effects of the 3-D bed form to the fluid

   The existence of both longitudinal and transverse instabilities of the coupled fluid-sediment system has been demonstrated by Vittori and Blondeaux (1992), suggesting a mechanism for the formation of at least one 3-D ripple type. However, these models will depend on parameterizations of the poorly known nearbed turbulence and sediment flux.

   The hypothesis that for directionally variable flows, bed forms become aligned in a direction such that the gross transport normal to the crest is maximized was confirmed in field experiments(Gallagher et al., 1998). The direction may differ substantially from the direction of the net bottom stress.

Even though we should notice that, at the present stage of research, considerable uncertainties are expected if untuned models are used to make absolute predictions for field conditions, after comparing 7 practical models with 5 field sites, Davies et al. (2002) pointed out that most of the formulas will converge for cases involving plane beds, with sand transport rates agreeing to well within an order of magnitude, and greatest divergence for cases involving rippled beds.

Since that we are not aiming to bridge the knowledge gaps in this study, the effects of the bed state module on the hydrodynamics module and the sediment transport module have been implemented in two ways. On the one hand, it is translated into the bed roughness, which is critical to define the flow/wave boundary layer, mixing layer, bed shear stress, the sediment transport rate, for the sediment transport module, and the bed boundary condition for the hydrodynamics module. On the other hand, that bed form module will describe the bed material pool for sediment transport module, which is termed as "bed bookkeeping" (refer to Section 3.2.1).

Based on Yalin's (1977) dimension analysis method, Van Rijn et al. (2004)'s bed roughness predictor is regarded as a good attempt and being used in this model:

$$\text{bed roughness} = f(\text{particle size, particle mobility, grain-related Reynolds number})$$

In hydraulically rough condition, grain-related Reynolds number is ignored suggested by Van Rijn and Walstra (2003).

The concept of an equivalent or effective sand roughness height k_s, introduced by Nikuradse Van Rijn et al. (2004), is used to simulate the hydraulic roughness of arbitrary roughness elements of the bottom boundary. The effective bed roughness (k_s) is the sum of: $k_{s,grain}$, grain related bed roughness; $k_{s,c}$, current related bed form bed roughness, which included the ripple, mega ripple, dune condition; $k_{s,w}$, wave related bed roughness; and k_a, apparent bed roughness (Van Rijn, 2007a).

In the lower flow regime the bed-load transport is strongly related to the migration of bed forms (ripples and dunes).

In the upper flow regime a thin high-concentration sheet flow layer is present just above the bed, in which the sediment concentrations vary from the order of 1,500 kg/m3 to about 10 kg/m3 over a thickness of the order of 0.01 m. This type of sediment motion is strongly related to particle-particle interaction, gravity, but not so much to turbulence-induced forces (largely damped due to the presence of particles (Van Rijn, 2007b), which is defined as bedload transport.

It needs to be noticed that the formula of Van Rijn (2007a, 2007b) for bed roughness prediction may not be correct from a purely physical point of view. The expressions are partly intuitive, engineering expressions rather than exact theoretical formulations in the sense that it provides the values of the right order of magnitude. Thus, it will be applied in the bed state module in this study.

2.3.2 Hydrodynamic modules

The hydrodynamic modules provide current velocity field and driving forces for sediment transport and geomorphological change, and subsequently, the effect of morphological change to flow field is taken into account. The hydrodynamic module used (but not necessarily) in current research is the existing sophisticated Delft3D system, such as Delft3D-FLOW, Delft3D-WAVE. The flow module solves unsteady shallow water equations in 2D depth average mode and 3D mode and provides flow velocity field. The wave module solves wave action balance equation and provides wave force, which enables wave-induced current. And the wave parameters are also provided to the sediment transport module to account for the stirring effect of wave motion on the sediments. Thus, bed level updating corresponds to the sediment (sand and mud) transport gradient following sediment mass conservation law.

The approach of coupling between the modules is also a crucial topic. Since the time scale for hydrodynamic processes is generally smaller than that of morphological processes, in the morphological module we need to extrapolate or accelerate the bed level updating. The state-of-the-art techniques are summarized in Roelvink (2006).

Hereafter, the governing equations for flow module and wave module are briefly overviewed first.

Governing equations for flow

In Cartesian coordinates and σ-layer in vertical, the 3D Shallow Water Equations read:

$$\frac{\partial \zeta}{\partial t} + \frac{\partial [HU]}{\partial x} + \frac{\partial [HV]}{\partial y} = Q \tag{2.15}$$

$$\frac{\partial U}{\partial t} + U\frac{\partial U}{\partial x} + v\frac{\partial U}{\partial y} + \frac{\omega}{H}\frac{\partial U}{\partial \sigma} - fV = -g\frac{\partial \zeta}{\partial x}$$

$$+\nu_H\left(\frac{\partial^2 U}{\partial x^2} + \frac{\partial^2 U}{\partial y^2}\right) + \frac{1}{H^2}\frac{\partial}{\partial \sigma}\left(\nu_V\frac{\partial U}{\partial \sigma}\right) + M_x \tag{2.16}$$

$$\frac{\partial V}{\partial t} + U\frac{\partial V}{\partial x} + V\frac{\partial V}{\partial y} + \frac{\omega}{H}\frac{\partial V}{\partial \sigma} + fU = g\frac{\partial \zeta}{\partial y}$$

$$+\nu_H\left(\frac{\partial^2 V}{\partial x^2} + \frac{\partial^2 V}{\partial y^2}\right) + \frac{1}{H^2}\frac{\partial}{\partial \sigma}\left(\nu_V\frac{\partial V}{\partial \sigma}\right) + M_y \tag{2.17}$$

where: the x-axis is chosen in the direction of the principal current ;
x, y, σ are coordination in σ layers system;
U and V are the Generalized Lagrangian Mean (GLM) velocity components in the x- and y-direction, which are sum of Eulerian velocity and Stokes' drift components

(Groeneweg, 1999; Walstra et al., 2000);

ζ: the free surface elevation;

H: water depth $H = (h + \zeta)$;

h: the bottom level with respect to the undisturbed water depth H;

M_x, M_y: external sink and source terms for momentum (by hydraulic structures, discharge or withdrawal of water, wave stress, etc.);

Q: the source or sink term of mass per unit area (discharge, withdrawal of water, evaporation, precipitation, etc.);

ν_H, ν_V : Horizontal and vertical kinematic viscosity coefficients (m^2/s).

Governing equations for waves

The governing equation for wave depends on different wave drivers. The most widely used drivers are spectral, short wave averaged and shortwave resolving (Roelvink & Reniers, 2012). Hereby only the general governing equation for the spectral approach is described.

The action density spectrum $N(\sigma, \theta)$ is used rather than the energy density spectrum $E(\sigma, \theta)$ since in the presence of currents, action density is conserved whereas energy density is not (Whitham, 1974). The independent variables are the relative frequency σ (as observed in a frame of reference moving with the current velocity) and the wave direction θ (the direction normal to the wave crest of each spectral component). The action density is equal to the energy density divided by the relative frequency: $N(\sigma, \theta)$ $= E(\sigma, \theta) / \sigma$.

In Cartesian co-ordinates (Holthuijsen, 2005):

$$\frac{\partial N}{\partial t} + \frac{\partial}{\partial x} c_x N + \frac{\partial}{\partial y} c_y N + \frac{\partial}{\partial \sigma} c_\sigma N + \frac{\partial}{\partial \theta} c_\theta N = \frac{S}{\sigma} \tag{2.18}$$

where $c_x, c_y, c_\sigma, c_\theta$ are the propagation velocities in x, y, σ, θ space (m/s).

The first term in the left-hand side of this equation represents the local rate of change of action density in time, and the second and third term represent propagation of action in geographical space. The fourth term represents shifting of the relative frequency due to variations in depths and currents. The fifth term represents refraction caused by bathymetry and current. The expressions for these propagation speeds are taken from linear wave theory (Dingemans, 1997; Mei, 1983; Whitham, 1974). The term at the right-hand side of the action balance equation is the source term in terms of energy density representing the effects of generation, dissipation and non-linear wave-wave interactions.

2.3.3 Sediment transport module

Overviews of sediment transport processes and the consequence bed roughness problem are given by Graf (1971), Vanoni (1975), Yalin (1977), for the river regime and by Sleath (1984), Nielsen (1992), Fredsøe and Deigaard (1992), Soulsby (1997), and Van Rijn (1993, 2006, 2007a, 2007b, 2007c, 2007d) for the coastal regime. In this research, also in Delft software system, the depth-integrated sediment transport is defined to consist of *bed-load transport*, which is the transport of sediment particles in a thin layer with thickness "close to the bed" of the order of centimeters; and *suspended load transport*, which is the transport of sediment particles above the bed-load layer. The suspended load transport can be determined by depth integration of the product of sediment concentration and fluid velocity from the top of the bed-load layer $z = z_a$ to the water surface. The total sediment transport is obtained as the sum of the net current contributed and wave contributed bed load $S_{b,c/w}$, and net current contributed and wave contributed suspended load $S_{s,c/w}$ transport rates, as follows:

$$S_{total} = S_{s,c} + S_{s,w} + S_{b,c} + S_{b,w} \qquad (2.19)$$

Hereafter, computional approaches and relevant processes for each component are briefly discussed.

Usually, the transport of particles by rolling, sliding, and saltating is called the bed-load transport.

In the past centuries, many studies have been carried out, such as:
1) Meyer-Peter and Mueller (MPM, (Meyer-Peter & Mueller, 1948)): Based on data analysis on flume experiments with both uniform and multiple sediment fractions , a relatively simple formula is proposed, which is still widely used nowadays.
2) Einstein (Einstein, 1950): statistical methods is introduced to represent the turbulent behavior of the flow. Einstein gave a detailed but complicated statistical description of the particle motion in which the exchange probability of a particle is related to the hydrodynamic lift force and particle weight. Einstein proposed the d_{35} as the effective diameter for particle mixtures and the d_{65} as the effective diameter for grain roughness.
3) Bagnold (Bagnold, 1966): Energy concept is introduced and sediment transport rate is related to the work done by the fluid.
4) Van Rijn (Van Rijn, 2007a, 2007b, 2007c, 2007d): Equations of motions of an individual bedload particle are solved. Saltation characteristics were computed. Particle traveling velocity and particle diameter are a function of the flow conditions. The bedload transport model for steady flow proposed is a parameterization of a detailed grain saltation model representing the basic forces acting on a bed-load particle, such as (Van Rijn, 2007a)'s simple recipe: the bedload transport magnitude may be roughly estimated as: $S_b = A(u - u_{cr})^B$, where A is a contant, and B is usually $2 \sim 5$.

However, a more dedicated approach is applied in this research. Bedload transport is calculated for all sand sediment fractions following the approach described by Van Rijn (2007b). In the first step, the magnitude and direction of the bed load sand transport are computed using the formulations with or without waves. The computed sediment transport vectors are then relocated from water level points to velocity points, by a numerical scheme, e.g., upwinding. Finally, the transport components are adjusted for bed-slope effects.

For bedload transport without wave ($S_{b,c}$), the magnitude on a horizontal bed is a function of: η, the relative availability of the sediment fraction in the mixing layer; u'_*, the effective bed shear velocity; D_*, the dimensionless particle diameter; T, the dimensionless bed-shear stress. u'_* and T are based on the computed velocity in the bottom computational layer. The direction of the bedload transport is parallel with the flow in the bottom computational layer.

For bed load and suspended load transport with wave ($S_{s,w}$ and $S_{b,w}$), the magnitude and direction on a horizontal bed are calculated using an approximation method developed by Van Rijn (2001). An estimate of the effects of wave orbital velocity asymmetry on sediment transport is done first. The magnitude of the bed load and suspended load transport is a function of Isobe and Horikawa (1982); Grasmeijer and Van Rijn (1998): d_{50}, the particle diameter; u_{cr} is the critical depth-averaged velocity for initiation of motion (based on a parameterization of the Shields curve); u_R is the magnitude of an equivalent depth-averaged velocity computed from the (Eulerian) velocity in the bottom computational layer, assuming a logarithmic velocity profile; U_{on} is the near-bed peak orbital velocity in onshore direction (in the direction on wave propagation) based on the significant wave height. The direction of the bedload transport vector is determined by the direction of the (Eulerian) near-bed current and the direction of wave propagation Van Rijn (2001); Soulsby (1997).

Suspended sediment transport rate by current ($S_{s,c}$) may be very roughly estimated as a function of representative median grain size, depth-averaged velocity and critical depth-averaged velocity (Van Rijn, 2007b).

However, for better estimation, suspended sediment transport rate ($S_{s,c}$) may be computed by a transport solver as for other passive constituents in the water column.

Suspended sediment transport equations

The 3D transport equation reads:

$$\frac{\partial hc}{\partial t} + \frac{\partial huc}{\partial x} + \frac{\partial hvc}{\partial y} + \frac{\partial wc}{\partial \sigma} = h[\frac{\partial}{\partial x}(D_H\frac{\partial c}{\partial x}) + \frac{\partial}{\partial y}(D_H\frac{\partial c}{\partial y})] + \frac{1}{h}\frac{\partial}{\partial \sigma}[D_V\frac{\partial c}{\partial \sigma}] + hS \quad (2.20)$$

where:

D_H, D_V : Horizontal and vertical diffusion coefficients (m^2/s);

ω : Vertical velocity of sediment particles in the σ-coordinate system. The settling velocity for sand is a function of particle diameter, water viscosity (Van Rijn, 1993). Vertical mixing coefficient of sediment is a function based on water depth, shear velocity etc. (Van Rijn, 1993). Reference concentration and reference height are calculated following Eq. 7.3.23-7.3.30 of Van Rijn (1993).

Water surface boundary

The vertical diffusive flux through the free surface is set to zero for suspended sediment.

$$-\omega_s c - D_V \frac{\partial c}{\partial z} = 0, \qquad at \ z = \zeta \tag{2.21}$$

where ζ is the free surface elevation.

Bed boundary

The exchange of material in suspension and the bed is modeled by calculating the sediment fluxes from the bottom computational layer to the bed, and vice versa. These fluxes are then applied to the bottom computational layer by means of a sediment source and/or sink terms in each computational cell. The sum of calculated fluxes for each sediment fraction is also applied to the bed in order to update the bed level. The boundary condition at the bed is given by:

$$-\omega_s c - D_{V,z_a} \frac{\partial c}{\partial z} = E - D, \qquad at \ z_a \ (water/bed \ interface) \tag{2.22}$$

where:

D is the sediment deposition rate of sediment fraction;

E is the sediment erosion rate of sediment fraction. The source and sink term are defined below.

When the concentration and velocity at the cell center are computed, the suspended sediment transport flux is given by definition:

$$S_{s,x} = \int_{z_a}^{\zeta} \left(uc - D_{H,x} \frac{\partial c}{\partial x} \right) \tag{2.23}$$

$$S_{s,y} = \int_{z_a}^{\zeta} \left(vc - D_{H,y} \frac{\partial c}{\partial y} \right) \tag{2.24}$$

Total transport

The total sediment transport is the sum of the changes due to suspended load, the suspended-load correction vector, and bedload. Estimation of each part is described above. And the process would be repeated for each sediment fraction. However, there is also a lumped way to estimate the total transport flux by many empirical formulas. They are widely used in many complex environments. For example, to estimate the time-averaged longshore net sediment transport rate in the surf zone, Bayram et al. (2001) compared the skill of six well-known formulas, i.e., formulas proposed by Bijker(1967), Engelund-Hansen(1967), Ackers-White(1973), Bailard-Inman(1981), Van Rijn (1984), and Watanabe (1992). All formulas were applied with standard coefficient values without calibration. He concluded that Van Rijn formula was found to yield the most reliable predictions over the range of swell and storm conditions covered by the field data set, while the Engelund-Hansen formula worked reasonably well, although with large scatter for the storm cases, and the Bailard-Inman formula systematically overestimated the swell cases and underestimated the storm cases. The formulas by Watanabe and Ackers-White produced satisfactory results for most cases, although the former overestimated the transport rates for swell cases and the latter yielded considerable scatter for storm cases. And the Bijker formula systematically overestimated the transport rates for all cases. Thus, in practice, re-calibration of the coefficient values by reference to field data is always helpful to improve their predictive capability (Spielmann et al., 2004).

2.3.4 Geomorphological bed level updating module

The bed level updating module is described as (Roelvink, 2006; Roelvink & Brøker, 1993):

$$(1-p) * \frac{\partial z_b}{\partial t} + \frac{\partial S_{total,x}}{\partial x} + \frac{\partial S_{total,y}}{\partial y} = Source/sink \tag{2.25}$$

where:
S_x, S_y represents the overall (bed load and suspended, induced by current and wave) sediment transport flux calculated above;
Source/sink includes only the dredging and dumping, and other sediment withdraw sources/sinks.

Roelvink (2006) listed different strategies of bed level updating, such as, tide-averaging method, or termed as MTM (refer to Section i and Figure 2.2), which consider the bottom fixed during the computing of hydrodynamics and sediment transport during a tidal cycle, and could be improved by continuity correction technique (Latteux, 1995; Cayocca, 2001); RAM (Rapid Assessment of Morphology) method, which related the

37

transportation to the water depth if assuming overall flow and wave patterns do not change for small bed level changes; online approach, accelerated by morphological factor (Lesser et al., 2004; Grunnet et al., 2004; Elias, 2006; Van De Wegen, 2010); and Parallel online approach by splitting the simulation into a number of parallel processes. By comparing the accuracy and efficiency, the online approach and parallel approach had advantages, which will be used in this research.

2.4 Development of geomorphological models: structured / unstructured grid solvers

Felippa (2004) classified the field of Mechanics into three major areas, i.e., theoretical, applied, and computational mechanics. Most computational fluid mechanics solvers for environmental studies treat water as a continuum; space is discretized using a structured or unstructured grid. Based on the discretization method, the dynamic system could be solved using finite element method (FEM), finite difference (FDM), boundary element method (BEM), finite volume method (FVM), spectral method and meshfree method. For solid mechanics problems, finite element methods currently dominate the field. However, for fluid mechanics, finite difference discretization methods are still important. Finite-volume methods, which address conservation laws, are important in passive transport. Spectral methods are based on transforms that map space and/or time dimensions to spaces where the problem is in continuous space. A recent newcomer is the mesh-free method, such as smooth particle hydrodynamics (SPH). Each numerical method has its own advantages and disadvantages. Hereafter, a list of widely used numerical solvers using structured grid and unstructured grid are very briefly overviewed below.

2.4.1 Structured grid: Delft3D

Delft3D-FLOW

Finite difference methods describe the unknowns x of the flow problem by means of point samples at the node points of a grid of co-ordinate lines. Truncated Taylor series expansions are often used to generate finite difference approximations of derivatives of x in terms of point samples of x at each grid point and its neighbors. Those derivatives appearing in the governing equations are replaced by finite differences yielding an algebraic equation for the values of x at each grid point. Smith (1985) had given a comprehensive overview of the finite difference method.

Flow module of Delft3D-FLOW is using finite difference method for structured grid. To discretize the 3D shallow water equations in space, the model area is covered by a rectangular or curvilinear grid in Cartesian (planar) or spherical (longitude and latitude) coordinates which is assumed to be orthogonal and well structured. Vertically users could also choose on σ-layer, or z-layer discretization. The artificial diffusive term introduced by the σ-layer discretization is eliminated following the method of Stelling and Van Kester (1994). Furthermore, horizontal pressure terms are simplified with Boussinesq approximation and horizontal Reynold's stress are simplified. The simplified form of the Navier-Stokes equations is solved with various types of boundary conditions, including solid walls, inflow and outflow, periodic boundaries, symmetry boundaries etc.

The variables are arranged in a pattern called the Arakawa C-grid (a staggered grid). In this arrangement, the water level points (pressure points) are defined in the center of a (continuity) cell; the velocity components are perpendicular to the grid cell faces where they are situated (see Figure 2.3). ADI (alternating direction implicit) method is used to solve the continuity and horizontal momentum equations (Leendertse, 1987). The advantage of the ADI method is that the implicitly integrated water levels and velocities are coupled along grid lines, leading to systems of equations with a small band width. Stelling (1984) extended the ADI method with a special approach for the horizontal advection terms. This approach uses two second-order discretization, a central discretization, and an upwind discretization to split the third-order upwind finite-difference scheme for the first derivative. The two second-order discretization are successively used in both stages of the ADI scheme. The scheme is denoted as a "cyclic method" (Stelling & Leendertse, 1991). This leads to a method that is computationally efficient, at least second-order accurate, and stable at Courant numbers of up to approximately 10. The diffusion tensor is redefined in the σ-coordinate system assuming that the horizontal length scale is much larger than the water depth (Mellor & Blumberg, 1985) and that the flow is of boundary layer type.

For spatial discretization of the horizontal advection terms, there are 3 options in Delft3D-FLOW: *WAQUA* (Stelling, 1984), *cyclic* (Stelling & Leendertse, 1991), *flooding* (Stelling & Duinmeijer, 2003). In *WAQUA* method, the third order upwind finite difference scheme is used for the horizontal advection terms. When normal advection term dominated, like in river case, the scheme has little dissipation and is used for the accurate prediction of water levels. For *cyclic* option, the horizontal advection terms are discretized into two second order scheme, a central scheme and an upwind scheme, and are applied alternatively for both half time steps. The resulting matrix is diagonally dominant and the iterative scheme converges fast. It is also a stable method with Courant number of up to 10 (for accuracy reason, value smaller than $4\sqrt{2}$ is recommended). The *flooding* scheme can be applied for problems that include rapidly varying flows for instance in hydraulic jumps and bores.

For all the validation cases in Chapter 4, Delft3D-FLOW is used as a reference model to compare with the newly developed morphology model. There all the cases in Delft3D-FLOW the *cyclic* option are used. For vertical advection and diffusion terms, implicit center scheme is employed.

Delft3D-WAVE SWAN

To simulate the evolution of wind-generated waves in coastal waters, the SWAN (Simulating WAves Nearshore) model is widely used and is employed in Delft software system. The SWAN model describes waves with the two-dimensional wave action density spectrum varying in time and space, accounting for refractive propagation of random, short-crested waves over arbitrary bathymetry and current fields (Booij et al., 1999; Ris et al., 1999). The processes of wave generation, whitecapping, nonlinear triad and quadruplet wave-wave interaction, bottom dissipation, depth-induced wave breaking and wave propagation through obstacles and wave-induced set-up of the mean sea surface can be computed explicitly (refer to Booij et al. (1999) for a more detailed description).

To model the energy dissipation in random waves due to depth-induced breaking, a spectral version of the bore-based model of Battjes and Jansen (1978) is used, here applied with a time-independent constant breaker parameter 0.73, and to model bottom-induced dissipation, the JONSWAP formulation (Hasselmann et al., 1973) is applied to compute bottom friction. The formulation for wave-induced bottom stress is modeled according to Fredsøe (1984).

An implicit second order upwind scheme in geographic space (x, y) and a mixed second order upwind/central scheme in spectral space (σ, θ) have been chosen for SWAN (Booij et al., 1999).

As a spectral model, SWAN does not attempt to represent physical processes at scales less than a wave length even in regions with very fine-scale mesh resolution. Phase resolving wave models should be employed at these scales if subwave length scale flow features need to be resolved. However, this fine-scale mesh resolution may be necessary for other reasons, such as representing the complex bathymetry and topography of the region, or to improve the numerical properties of the computed solution. This notice also keeps for the unstructured SWAN.

Delft3D-MOR

Suspended sediment transport

The transport equation in Delft software system is formulated in a conservative form (finite-volume approximation) and is solved using the so-called "cyclic method" (Stelling & Leendertse, 1991). For steep bottom slopes in combination with vertical stratification, horizontal diffusion along σ-planes introduces artificial vertical diffusion (Huang & Spaulding, 1996). Stelling and Van Kester (1994) purposed an algorithm to approximate the horizontal diffusion along z-planes in a σ-coordinate framework. In addition, a horizontal Forester filter (Forester, 1979) based on diffusion along σ-planes is applied to remove any negative concentration values that may occur. The Forester filter is mass conserving and does not cause significant amplitude losses in sharply peaked solutions.

Bedload sediment transport

Notice that all the variables are located at the grid cell center. Therefore, it is necessary to integrate the sediment transport vectors to the velocity points. In most general way, the upwinding algorithm is applied.

The bedload transport vector components are computed at the water level points in the staggered grid ($S_{b,u}$ in Figure 2.3), as are the suspended-sediment sources and sinks (refer Section 2.3.3). The bedload vector components at the velocity points ($S_{b,uu}$), around the perimeter of each cell control volume, are determined by transferring the appropriate vector components from the adjacent water level point half a grid cell upwind. In the u direction, the transport at the u-velocity point, $S_{b,u}^{(m,n)}$ is given to $S_{b,uu}^{(m,n)}$ which denotes the u-component of the transport computed at the upwind water level point. In the v direction, the transport at the v-velocity point, $S_{b,v}^{(m,n+1)}$ is given to $S_{b,vv}^{(m,n)}$ because the bedload transport direction opposes the grid direction (refer Figure 2.3).

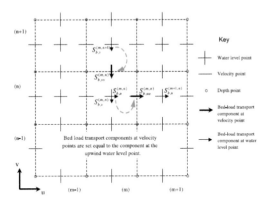

Figure 2.3: The upwinding scheme for the bedload transport component at velocity points (Lesser et al., 2004) (Note that the control volume of sediment transport located at the water level point).

This upwind shift ensures numerical stability and allows the implementation of an extremely simple morphological updating scheme. However, if the vector directions in adjacent water level points oppose, then a central scheme is used.

2.4.2   Unstructured grid: Delft3D D-Flow Flexible mesh

D-Flow FM follows the numerical concepts of Delft3D and SOBEK1D2D (Kernkamp et al., 2011). Its grid allows polygon-shaped cells of wide range degree. The unstructured grid approach is based on the combination of 2D/3D finite-volume cells with 1D flow networks into a single grid. Similar to Delft3D structured model, a staggered grid is used.

And in the model, 1D, 2D or 3D modelling concepts are combined. For instance, river tributaries can now easily be modelled as 1D branch. Furthermore, it is based on an efficient matrix solver which combines Gaussian elimination (to decrease the degree to minimum) and conjugate gradients (CG). Time integration is done with the implicit θ-method. The advection terms are treated explicitly.

2.4.3   Unstructured grid: SWAN

The unstructured-mesh version of SWAN implements an analog to the four-direction Gauss-Seidel iteration technique employed in the structured version to assure unconditional stability ("Computation of wind-wave spectra in coastal waters with SWAN on unstructured grids", 2010). SWAN computes the wave action density spectrum $N(\sigma, \theta)$ at the vertices of an unstructured triangular mesh, and it orders the mesh vertices so it can sweep through them and update the action density using information from neighboring vertices. It then sweeps through the mesh in opposite directions until the wave energy has propagated sufficiently through geographical space in all directions.

2.4.4   Unstructured grid: ADCIRC

In ADCIRC, the solution is implemented using Lagrange linear finite elements in space and three- and two-time-stage schemes in time for the equations, respectively (Westerink et al., 2008; Dietrich et al., 2011). An implicit time discretization is applied for all linear terms and Coriolis, atmospheric-pressure-forcing, and tidal-potential terms being treated explicitly and an explicit discretization is used for other nonlinear terms, which leads to a Courant restriction (below 1). Coupling model of ADCIRC and Unstruc SWAN is applied to study storm surge and hurricane waves (Dietrich et al., 2011), which gives a good example for models coupling and both models use triangular grid.

Kubatko et al. (2006) developed a unstructured grid two-dimensional, depth-integrated morphodynamic model by online coupling the ADCIRC, the continuous Galerkin (CG) finite element hydrodynamic model and a sediment transport/bed evolution model using discontinuous Galerkin (DG) method for the solution of the sediment continuity equation.

It incorporated upwinded numerical fluxes and slope limiters to provide sharp resolution of steep bathymetric gradients that may form in the solution, and a local conservation property that conserves sediment mass on elemental level is assured.

2.4.5   Unstructured grid: TELEMAC

Detailed hydrodynamics of TELEMAC refer to Hervouet (2007). TELEMAC uses finite element methods based on simple piecewise functions (e.g. linear or quadratic) valid on elements to describe the local variations of unknown flow variables x. The governing equation is precisely satisfied by the exact solution x. If the piecewise approximating functions for x are substituted into the equation it will not hold exactly and a residual is defined to measure the errors. Next the residuals (and hence the errors) are minimized in some sense by multiplying them by a set of weighting functions and integrating. As a result we obtain a set of algebraic equations for the unknown coefficients of the approximating functions. TELEMAC solves the non-conservative momentum equations and the conservative depth-integrated continuity equation. All variables are located in the vertices (nodes) of an unstructured grid (triangles for 2D, prisms for 3D) (Wenneker, 2003).

Atkinson et al. (2004) reviewed the similarities between the quasi-bubble used by TELEMAC and the generalized wave continuity equation solutions (ADCIRC) to the shallow water equations and concluded that these schemes are very similar, even identical if the weighting parameters are properly selected in the linear finite element approximation in ADCIRC.

A successful example of the integration of a finite volume solver (Delft-WAQ) and finite element solver (TELEMAC) can be found in Postma and Hervouet (2007).Postma and Hervouet (2007) formulated a finite volume analogue of a finite element schematization of transport equation with deliberated processing of integration area, boundary nodes to make sure the nodal and element-wise mass conservative, which shows the potential consistent and mass conserving coupling for the finite element TELEMAC model for 3D flow dynamics with finite volume Delft3D-WAQ model for water quality (and particle transportation).

For morphology updating, a two step approach is used in SISYPHE. Firstly a prediction of sediment transport flux is made followed by a correction step by θ averaging to smooth out the wiggles to some extend, which makes the SISYPHE accurate during the beginning period.

2.4.6   Unstructured grid: FVCOM

FVCOM (Chen-C et al., 2003) is a 3D unstructured grid, finite-volume coastal ocean model. FVCOM discretizes the integral form of the governing equations. These integral equations are solved numerically by flux calculation used in the finite difference method over triangular grids, but still with the finite volume approach along the cell interface to guarantee mass conservation in both individual control elements and the entire computational domain. In numerical aspects, mode splitting" is used, i.e., the currents are divided into external and internal modes that can be computed using two different time steps. For the external 2D mode, the fourth-order Runge-Kutta time-stepping scheme with second-order accuracy is used. For the internal 3D mode, the momentum equations are solved using a combined explicit and implicit scheme in which the local change of the currents is integrated using the first-order accuracy upwind scheme. The advection terms are computed explicitly by a second-order accuracy Runge-Kutta time-stepping scheme.

Sediment transport in the FVCOM is also an example of coupling for unstructured grid. The suspended sediment transport solver of FVCOM is coupled with CSTM/ROMS ("Development of a three-dimensional, regional, coupled wave, current, and sediment-transport model", 2008). Functionally CSTM/ROMS is quite similar to Delft3D-SED and Delft3D-MOR in morphodynamic aspects even though CSTM/ROMS seems less sophisticated. Only the Meyer-Peter Meller (1948) formulation for unidirectional flow and the formulae of Soulsby and Damgaard (2005) for flow and wave situation are used for bedload in CSTM/ROMS.

2.4.7   Structured grid: XBeach

XBeach is a two-dimensional open source model for wave propagation, long waves and mean flow, sediment transport and morphological changes of the nearshore area, beaches, dunes and back barrier during storms ("Modelling storm impacts on beaches, dunes and barrier islands", 2009). XBeach also uses a rectangular or curvilinear staggered grid. The bed levels, water levels, water depths and concentrations are defined in cell centers, and velocities and sediment transports are defined at the cell interfaces. In the wave energy balance, the energy, roller energy and radiation stress are defined at the cell centers, whereas the radiation stress gradients are defined at the cell interfaces.

Time integration is done with first-order upwind explicit scheme with an automatic time step is applied for flow. Momentum-conserving form is used which is especially suitable for drying and flooding and which allows a combination of sub- and supercritical flows (flooding scheme in Delft3D-FLOW as described in Stelling and Duinmeijer (2003)). Numerical stability is set to be at the top priority, thus first-order accuracy of upwind scheme is accepted, since it is need for small space steps and time steps anyway, to represent the strong gradients in space and time in the nearshore swash zone ("Modelling storm impacts on beaches, dunes and barrier islands", 2009).

2.4.8   Conclusions

There are more morphodynamic model using structured grid, such as, CSTM/ROMS ("Development of a three-dimensional, regional, coupled wave, current, and sediment-transport model", 2008), and models using unstructured grid, such as, FINEL, Delfin (Ham et al. 2005), SLIM (http://www.climate.be/SLIM), and Mike system including both structured grid and triangular grid etc.. These models are proved to be applicable for specific practical applications if the modelers are properly trained. However, there is not one system yet that could have the combination where the grid allows polygon-shaped cells of arbitrary degree for 1D, 2D or 3D modeling concepts, which is getting increasingly demanded for practical applications. The model D-Flow Flexible mesh, if completed, would be a good candidate for a more flexible mixture of 1D, 2D and 3D flow modeling capabilities, while the generic morphological model to be developed in this study would offer the morphology modeling capabilities on a combination of triangles and quadrilaterals.

2.5   Summary

The coastal system is studied with the help of different types of models, such as, coastline model, profile model, area model and local model to describe the processes behind in various spatial and temporal scales. With the new technologies developing, some new methods, such as, data driven modeling, data assimilation techniques, non-linear modeling are developed. The performance of different models can be evaluated by a number of statistical tests, bias, accuracy and skill.

Brief overview in this chapter showed that still many gaps are to be bridged after many years of research due to the complex nature of the coastal system. Integrated modeling studies increasingly demand a flexible generic morphodynamics model of 1D, 2D and 3D modeling capabilities on unstructured grid. The proposed generic morphodynamics model described in the next chapter aims to that.

In this research, the proposed generic morphodynamics model will consist of 4 new or extended modules in addition to the existing hydrodynamics modules (Section 2.3.2): the bed state module (Section 3.2), which acts as the bed roughness predictor and the bed material bookkeeping system; the sediment transport module (Section 3.3), providing the sediment transport vector; the bed level change module (Section 3.4); and a vegetation population dynamic model described in Chapter 5.

Chapter 3

Developement of the Generic

Coastal Morphological Model

3.1 Introduction

In addition to the existing hydrodynamic modules described in Section 2.3, newly developed or extended components of the generic coastal morphological model are described in this chapter. The generic coastal morphological model consists of six modules in total. Deltares is developing the flexible flow module D-Flow FM which can be coupled to structured or unstructured SWAN. Since this flow module hasn't been completed yet, we have used the existing modules Delft3D-FLOW and WAVE (refer to Section 2.4.1, (Deltares, 2010a) and (Deltares, 2010c)) instead. Three new modules (sediment transport, bed state module, bed update module) have been implemented in Delft3D-WAQ partly reusing old code of very elementary sediment transport functionality in WAQ and the conceptal simulation code in Delft3D online MOR.

Especially, the morphodynamic modeling technique for unstructured grid is addressed in Section 3.4. Following that, the velocity integration algorithm is discussed in Section 3.5. In the end, the model structure and dataflow are demonstrated, and then the substance definitions and corresponding processes are listed in Section 3.6. The communication approaches among different modules are also discussed.

Special emphasis is given to the numerical schemes adaptive to both structured and unstructured grid and the open model structure which make this morphological model to be a generic model.

3.2 Bed state description module

The bed state description module includes a bed form predictor and a bed state book keeping system.

Bed state affects hydrodynamics and sediment transport in two ways. First, it impacts on the bed roughness, which is critical to define the flow/wave boundary layer, mixing layer, bed shear stress, the sediment transport rate for the sediment transport module, and the bed boundary condition for the hydrodynamics module. Second, bed state is described the final results of the bed material after the sediment transport and morphological change, which is termed as "bed state book keeping" (refer to Section 3.2.1).

3.2.1 Bed state book keeping system

Book keeping of the sediment fractions in different bed layers and their interactions are implemented, which make it possible to simulate the graded sediment processes. The similar layering approach has been applied in SOBEK-GRADED, by subdivision of bed materials into fractions (Ribberink, 1987), and also in Delwaq-G, by considering of nutrients exchange between sediment-water. Nevertheless, neither of these approaches considers sediment mass exchange between the bed and water column when sediment transport and the morphological changes is not negligible.

In this module, the bed consists of various sediment layers. The bookkeeping system tracks the sediment composition and the state of each layer in the computational volume for all sediment layers.

Briefly, vertically the sediment can be exchanged between water body and transport layer (entrainment and deposition), and the sediment can be packed into, and can be replenished from the exchange layer (consolidation and liquefaction). The underlying layers are sediment buffers for the upper layers. Horizontal transport only happens in the transport layer, which is termed as bed load transport. Thus, the thickness of the transport layer is an analogy of Van Rijns reference height (Van Rijn et al., 2004).

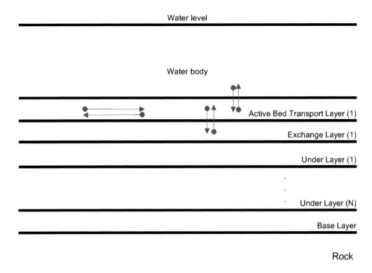

Figure 3.1: Sketch of the book keeping system

3.2.2 Bed form predictor

The principle of the bed form predictor in this study follows Van Rijn (2007a, b, c, d). The bed form is represented by the bed roughness, which is critical to define the flow/wave boundary layer, mixing layer, bed shear stress, the sediment transport rate, for the sediment transport module, and the bed boundary condition for the hydrodynamics module.

Van Rijn et al. (2004) proposed a bed roughness predictor based on Yalin's (1977) dimension analysis method:

$bed\ roughness = f$(particle size, particle mobility, grain-related Reynolds number)

Nikuradse introduced the concept of an equivalent or effective sand roughness height k_s to simulate the hydraulic roughness of arbitrary roughness elements of the bottom boundary. The effective bed roughness (k_s):

$$k_s = k_{s,grain} + k_{s,w} + k_{s,c} + k_a \tag{3.1}$$

where:

$k_{s,c}$: current related bed roughness , which included the ripple, megaripple, dune condition;

$k_{s,w}$: wave related bed roughness;

49

k_a: apparent bed roughness(Van Rijn, 2007a; Soulsby, 1997).

It is realized that the formula may not be correct from a purely physical point of view. The expressions are partly intuitive, engineering expressions rather than exact theoretical formulations in the sense that it provides the values of the right order of magnitude. Thus it is applied in this model and updated with the bed book keeping system.

3.3   Sediment transport module

Sediment transport equation and the implementation of sediment transport in structured grid has been briefly discussed in Chapter 2 (Section 2.3.3 and Section 2.4.1). Hereby, one feature to be addressed is the cohesive sediment transport is integrated in the same structure as non-cohesive sediment transport.

In this model, in each unstructured cell, for non-cohesive sand, source and sink terms are defined as that for structured grid described in Lesser et al. (2004):

$$E = \alpha_2 c_a \left(\frac{D_V}{\Delta z} \right) \tag{3.2}$$

$$D = \left[\alpha_2 \left(\frac{D_V}{\Delta z} \right) + \alpha_1 \omega_s \right] c_{kmx} \tag{3.3}$$

Here, α_1 is a factor to correct the concentration at the reference layer centre to the bottom at the reference layer, α_2 is a factor to express the concentration gradient at the bottom of the reference layer as a function of the reference concentrations c_a and c_{kmx}.

Additionally, for mud fractions in each unstructured cell, the source and sink terms are defined based on the Partheniades-Krone formulations:

$$E = M \left(\frac{\tau_{c,w}}{\tau_{cr,e}} - 1 \right) \quad when \quad \tau_{c,w} > \tau_{cr,e}$$

$$= 0.0 \quad when \quad \tau_{c,w} < \tau_{cr,e} \tag{3.4}$$

$$D = \omega_s c_b \left(1 - \frac{\tau_{c,w}}{\tau_{cr,d}} \right) \quad when \quad \tau_{c,w} > \tau_{cr,d}$$

$$= 0.0 \quad when \quad \tau_{c,w} < \tau_{cr,d} \tag{3.5}$$

where:

$$c_b = c\left(z = \frac{\Delta z_b}{2}, t\right) \tag{3.6}$$

Furthermore, for mud simulation, the settling velocity is modified based on the salinity, while the vertical mixing coefficient for sediment is set to be equal to the vertical fluid mixing coefficient. For mud sediment fractions, the source and sink terms are always assigned in the bottom computational cell rather than the reference layer and are computed with the Parthenaides and Krone formulations (Partheniades, 1965, 2009).

Once the source and sink term are defined, the Eq.2.20 could be solved numerically in the Delft3D-WAQ system where there are about 22 numerical schemes and advection-diffusion solvers available to use. Scheme 15 and Scheme 19, 20 (also 21, 22) are recommended. Scheme 15 is a iteration solver (for unstructured grid). It is an implicit upwind scheme in both horizontal and vertical directions. So it is unconditionally stable with first order accuracy in both space and time. GMRES solver is applied with Gauss-Seidel pre-conditioner in horizontal. Direct LDU solver is applied in vertical. Also the scheme 21 and 22 could be used for unstructured grid. They are extended from Scheme 15/16 with flux correction transport (FCT) technique, which correct the oscillations in each time step to fulfill the TVD criterion (refer to 3.4.3 for detail).

Scheme 19, 20 are ADI scheme (for structured grid) similar to the solvers in Delft3D-FLOW, the horizontal advection terms in the scalar transport equation are approximated by the sum of a second order upwind scheme and a second-order central scheme. The horizontal diffusion terms are computed by Crank-Nicholson scheme explicitly (Stelling & Leendertse, 1991). The difference of Scheme 19 and 20 is that in vertical center scheme is used in Scheme 19 (with Forester filter to make sure positive), and upwind scheme is used in Scheme 20 (Deltares, 2010b).

Bedslope effects on sediment transport is included in this model. As bedload transport is continuously attached with the bed, the bed slope affects the magnitude and direction of the bedload transport vector. A longitudinal slope in the direction of the bedload transport modifies the magnitude of the bedload vector as follows (modified from (Bagnold, 1966)):

$$S_{b,uu} = \alpha_s S_{b,uu}, \quad S_{b,vv} = \alpha_s S_{b,vv} \tag{3.7}$$

where

$$\alpha_s = 1 + f_{alfabs} * \left[\frac{tan\phi}{\cos(\tan^{-1}(\frac{\partial z}{\partial s}))(\tan\phi - \frac{\partial z}{\partial s})} - 1\right] \tag{3.8}$$

in which f_{alfabs} is a tuning parameter, $\frac{\partial z}{\partial s}$ is the bed slope in the direction of the bedload transport (positive down), ϕ is the internal friction angle of bed material (assumed to be 30°).

A transverse bed slope also modifies the direction of the bedload transport vector. This modification is broadly based on the work of Ikeda (1982) and is computed as follows:

$$S_{b,uu} = S_{b,uu} - \alpha_n S_{b,vv}, \quad S_{b,vv} = S_{b,vv} - \alpha_n S_{b,uu} \tag{3.9}$$

where

$$\alpha_n = f_{alfabn} \left(\frac{\tau_{b,cr}}{\tau_{b,cw}}\right)^{0.5} \frac{\partial z}{\partial n} \tag{3.10}$$

in which f_{alfabn} is a tuning parameter, $\tau_{b,cr}$ is the critical bed shear stress, $\tau_{b,cw}$ is the bed shear stress due to current and waves, $\frac{\partial z}{\partial n}$ is the bed slope normal to the unadjusted bedload transport vector.

3.4   Morphodynamic module

3.4.1   Sediment mass conservation and bed level updating

The governing equation of the bed level updating module reads (Section 2.3.4):

$$(1 - p) * \frac{\partial z_b}{\partial t} + \frac{\partial S_{total,x}}{\partial x} + \frac{\partial S_{total,y}}{\partial y} = Source/sink \tag{3.11}$$

Choices of various numerical schemes are provided for the bed level updating equation in this modeling system. The schemes are capable for unstructured grid. First order upwind scheme is the default option because of its good stability and simplicity.

Roelvink (2006) reviewed different strategies of bed level updating, such as, tide-averaging approach, rapid assessment of morphology approach, online approach, and parallel online approach. For the sake of both accuracy and efficiency, the online approach is used in this study. The bed level is updated and communicated simultaneously with other modules during the computation. The bed updating algorithm for curvilinear/structured grid has been discussed in Section 2.4.1 and hereafter the bed updating algorithm for unstructured grid is focused on.

3.4.2   Bed updating algorithm for unstructured grid

As the algorithm for structured grid, the flow module provides flow field information and Delft3D-WAQ will get the information through coupling files (online or specific coupling program). The representative integrated velocity vector at the grid cell (segment, element) centre is computed by the velocity integration scheme (refer to Section 3.5). The volumetric (bedload) sediment transport fluxes (vector) at the grid cell (center) are computed consequently, which call various sediment transport formulae, like Van Rijn1993, Van Rijn2004, EH, etc. The suspended sediment transport fluxes are simulated through the entrainment and deposition flux in each grid cell (center).

The volumetric (bedload) sediment transportation rate (vector) (m^2/s) at the Delft3D-WAQ exchange is computed by upwinding/averaging the sediment transport fluxes from the adjacent grid cells. The bed updating algorithm is described as: to compare the normal components of adjacent cells, and then to upwind/average the total vector to the exchange.

The following scheme carries out the upwinding/averaging processes:

- Let S_{n1} be the normal component at the cell exchange of the sediment transport vector S_1 of the "from" segment (grid cell) (see Eq. (4.18));

- let S_{n2} be the normal component at the cell exchange of the sediment transport vector S_2 of the "to" segment (grid cell) (see Eq. (4.20));

- if $S_{n1} > 0$ and $S_{n2} > 0$, use S_1;

- if $S_{n1} < 0$ and $S_{n2} < 0$, use S_2;

- otherwise use $(S_1+S_2)/2$.

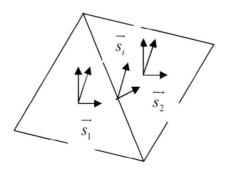

Figure 3.2: The sediment transport vector at the grid cell (segment) center and at the exchange

If $\vec{S}_1$ and $\vec{S}_2$ are the sediment transport rate vector at the cell center, and $\vec{n}_i$ is the unit vector vertical to the normal direction of exchange, $\vec{S}_i$ (m^2/s) is the sediment transport rate vector at the exchange:

$$\vec{S}_1 = S_{1,x}\vec{x} + S_{1,y}\vec{y} \tag{3.12}$$

$$S_{1,n_1} = \vec{S}_1 \cdot \vec{n}_1 = S_{1,x}(\vec{n}_1 \cdot \vec{x}) + S_{1,y}(\vec{n}_1 \cdot \vec{y}) \tag{3.13}$$

where:

$$(\vec{n}_1 \cdot \vec{x}) = cos(\phi_i) \tag{3.14}$$

$$(\vec{n}_1 \cdot \vec{y}) = sin(\phi_i) \tag{3.15}$$

$$\tag{3.16}$$

Then:

$$S_{1,n_1} = S_{1,x}cos(\phi_i) + S_{1,y}sin(\phi_i) \tag{3.17}$$

Similarly:

$$\vec{S}_2 = S_{2,x}\vec{x} + S_{2,y}\vec{y} \tag{3.18}$$

$$S_{2,n_1} = \vec{S}_2 \cdot \vec{n}_1 = S_{2,x}(\vec{n}_1 \cdot \vec{x}) + S_{2,y}(\vec{n}_1 \cdot \vec{y}) \tag{3.19}$$

Then:

$$S_{2,n_1} = S_{2,x}cos(\phi_i) + S_{2,y}sin(\phi_i) \tag{3.20}$$

If upwinding (first order accuracy):

$$if \ (S_{1,n_1}) > 0 \cap (S_{2,n_1}) > 0 \ then \ \vec{S}_i = \vec{S}_1 = (S_{1,x}\vec{x} + S_{1,y}\vec{y}) = (S_{i,x}\vec{x} + S_{i,y}\vec{y}) \tag{3.21}$$

$$if \ (S_{1,n_1}) < 0 \cap (S_{2,n_1}) < 0 \ then \ \vec{S}_i = \vec{S}_2 = (S_{2,x}\vec{x} + S_{2,y}\vec{y}) = (S_{i,x}\vec{x} + S_{i,y}\vec{y}) \tag{3.22}$$

Otherwise centralization (second order accuracy):

$$\vec{S}_i = \frac{1}{2}(\vec{S}_1 + \vec{S}_2) = \frac{1}{2}(S_{1,x}\vec{x} + S_{1,y}\vec{y}) + \frac{1}{2}(S_{2,x}\vec{x} + S_{2,y}\vec{y}) \tag{3.23}$$

and then:

$$S_{i,n} = \vec{S}_i \cdot \vec{n}_i = S_{i,x}(\vec{n}_i \cdot \vec{x}) + S_{i,y}(\vec{n}_i \cdot \vec{y}) = S_{i,x}cos(\phi_i) + S_{i,y}sin(\phi_i) \tag{3.24}$$

By multiplying the sediment transport rate at the exchange with the width of exchange, the total bed load sediment discharge (flux) is computed and is distributed to the two neighboring cells. Rate of change of thickness sand layer (flux) for each cell is also computed simultaneously and managed by the bed state book keeping system (refer to Section 3.2.1).

Therefore, the sediment transport rate through exchange (cellface) Q (m^3/s) is:

$$Q_{i,s} = S_{i,n} \cdot width_i \tag{3.25}$$

The sum of the total sediment transport flux, the suspended sediment entrainment and deposition flux, with suspended sediment transport correction and bedslope correction and some other corrections, and dumping/dredging flux are used to compute the depth change of cell centers (refer to Eq. 3.11):

$$\Delta z_{1,2} = \frac{\Delta t \left(\sum_{interface_i} Q_{i,s} \right)}{Area_{cell}} \tag{3.26}$$

When using the unique value of sediment transport discharge in exchange for both cells, the sediment mass conservation is automatically assured.

3.4.3 Discussion on stability and accuracy of numerical schemes

To solve the bed mass conservation equation using accurate and stable numerical schemes is crucial in the whole morphological modeling system. Minimized errors in modules always lead to a more accurate overall system. Moreover, when morphological factor is applied, the small errors might be amplified. Furthermore, to model the micro morphology bed forms in high spatial resolution, like sand waves, needs proper accuracy. The macro morphology bed forms, like sand dunes, might introduce a kind of discontinuity in the bed level, where the stability of the numerical solvers gains more concerns. Thus the stability and accuracy of numerical schemes are discussed here.

The most common choice is the upwind scheme, as we chose for Delft3D structured grid. It is robust and easy to implement for both structured grid and unstructured grid. It is automatically fulfill the TVD property. Unfortunately it is has only first order accuracy and suffers from diffusion. Second order center scheme is then used, but it suffers from oscillations and instability. Johnson and Zyserman (2002) mentioned that there might some nature processes or implementations might help to limit the oscillations, like bedslope correction, including random wave to smooth out the sediment transport gradients, or assumption of equilibrium suspended sediment transport. Furthermore, he suggested a Lax-Wendroff type scheme, which the disadvantages are:

- If Courant number is less 1, the scheme is stable and no damping, but need 7 grid points or more. The implementation may be still possible for structured grid, but for unstructured grid, it would be too complex;

- If Courant number is close to 1, there will be oscillations. If the oscillations are limited in the 7 grid points, then it would damp. But if the oscillations are in the whole area, then the amplification factor of the spatial oscillations does not reduce monotonically with Courant number (Johnson & Zyserman, 2002). Extra smoothing is then necessary.

Callaghan et al. (2006) suggested a general non-oscillating scheme by using two staggered grids, moving the solution from one grid to another with each time step. This approach is rather complex to apply in unstructured grid as well. Hudson et al. (2005) elaborated more on the flux limiters to fulfill the TVD criterion (discussed below). The importance of the source terms discretization in the bed mass conservation equation is also discussed. But his approach is even only in 1D. Also for structured grid, Long et al. (2008) introduced a fifth order Euler-WENO (forward time, back space, weighted essentially non-oscillatory) scheme to solve bed mass conservation equation. This scheme is a shock capturing scheme. With TVD correction, it has significant advantages by having artificial viscosity and filtering processes, and good for phase-resolving sediment transport models. The disadvantage may be that it needs iteration in solving the new bed level and sediment transport flux and also only validate for structured grid.

In the newly developed morphological model developed in this study, 3 options are available, i.e., upwind scheme, central scheme and second order upwind scheme with TVD filter. Hereafter brief introduction on the simplification of governing equation and the analytical solutions are given. Then the implementation in this generic morphological model is described. The numerical techniques used here are also possible to apply in other morphological models.

Governing equation and analytical solution

The simplified 1D governing equation for morphological updating reads:

$$\frac{\partial h}{\partial t} + \frac{\partial S}{\partial x} = 0 \tag{3.27}$$

where h is the bed level, S is the sediment transport flux for a given porosity. Assume S is a function of h, i.e. $S \sim h^b$, the equation 3.27 could be easily transformed into a general homogenous hyperbolic partial differential equation (refer to Section 4.2.1 for detailed derivation):

$$\frac{\partial h}{\partial t} + c \frac{\partial h}{\partial x} = 0 \tag{3.28}$$

where $c = b\frac{S}{h}$. The value of b usually is 5 (Engelund and Hansen formula) or 3 in coastal regions (Nielsen, 1992). For simplicity, c is assumed to be constant hereby, thus analytical solution of this linear equation and the dispersion relation are:

$$h = h_0 e^{i(\omega t - kx)} \tag{3.29}$$
$$\omega = ck \tag{3.30}$$

First order upwind scheme

Discretized using the first order upwind scheme:

$$h_i^{n+1} = h_i^n - c\frac{\Delta t}{\Delta x}(h_i^n - h_{i-1}^n) \tag{3.31}$$

Here the subscript i denotes the spatial step, and the superscript n denotes the time step.

Expanding with Taylor series in time and space and rearrangement, the equation 3.28 is solved as:

$$\frac{\partial h}{\partial t} + c\frac{\partial h}{\partial x} = \alpha\frac{\partial^2 h}{\partial x^2} + H.O.T. \tag{3.32}$$

where $\alpha = \frac{c}{2}(\Delta x - c\Delta t)$.

Performing a von Neumann stability analysis on the scheme, the analytical solution and the dispersion relation become then:

$$h = h_0 e^{-\alpha k^2 t} e^{i(kct - kx)} \qquad (3.33)$$

$$\omega = ck + i\alpha k^2 \qquad (3.34)$$

When CFL number $c\Delta t/\Delta x < 1$, $\alpha > 0$, the upwind scheme is stable, and also h is less than h_0, but results always stay positive. Since for long wave, $k \to 0$, $e^{-\alpha k^2 t} \to 1$, longer waves tend to remain and short waves are to vanish. The wave phase celerity is accurate, while the amplitude is damped. Thus there are no spurious oscillations generated at the sharp edge of the wave form, but the scheme suffers from the spurious diffusion due to the introduction of the artificial diffusion term. The truncation error shows that the upwind scheme gives first order accuracy in time and in space. If $c\Delta t/\Delta x = 1$, then the upwind scheme is not only stable, but also $h = h_0$, without damping. However, in practical problems, using curvilinear grid, it is difficult to set uniform celerity, thus space step and time step to assure the morphological courant number to be 1.

Second order central scheme

Central difference approximations have null dissipation error in a linear setting, and therefore may be ideal candidates for computations free of spurious numerical diffusivity.

Expanding with Taylor series in time and space and rearrangement, the equation 3.28 is solved as:

$$\frac{\partial h}{\partial t} + c\frac{\partial h}{\partial x} = \beta_1 \frac{\partial^3 h}{\partial x^3} + \beta_2 \frac{\partial^2 h}{\partial t^2} + H.O.T. \qquad (3.35)$$

where $\beta_1 = \frac{c}{6}(-\Delta x^2)$ and $\beta_2 = (-\Delta t^2)$. Thus this scheme is in second order accuracy in space and first order in time.

Central derivative approximations have been widely used in the literature, especially for wave propagation problems where nonlinearities are weak. However, it is known that application of standard central discretization leads to numerical instability, owing to accumulation of the aliasing errors resulting from discrete evaluation of the nonlinear convective terms (Phillips, 1959). It provides second order accuracy, but it is unconditionally unstable. This scheme could only be used in a small range of applications.

Total variation diminishing (TVD) criterion

As stated above, the solution of some higher order schemes, e.g., Eq. 3.35, might cause oscillation with different wave length. A mathematical criterion total variation diminishing (TVD) is introduced to indicate that no new maxima and minima are generated higher or lower than those that were already in the solution. The first order schemes always obey this criterion. In the morphological modeling case, it is:

$$\Sigma_i |h_{i+1}^{n+1} - h_i^{n+1}| \leq \Sigma_i |h_{i+1}^n - h_i^n| \tag{3.36}$$

In this study, the TVD criterion is activated optionally to limit sharp slope in bed levels.

Other higher order schemes

Higher order schemes might be favorable for higher accuracy in solving the bed update equations numerically. However, higher order schemes always request more computational effort. Furthermore, due to the complex environment, shock waves in the bed forms either hardly appear in the seabed, or are not resolved by the measurements. In modeling practice, the schemes mentioned above already give out good results. Thus higher order schemes only remain good research topics rather than applications in engineering practice. It is interesting rather than important; therefore, other higher order schemes are not included in this generic morphological modeling system.

3.4.4   Discussion on mass conservation

There are concerns that the mass conservation law in the modeling system might be violated due to bed update and morphological change, which might become worse when morphological acceleration factor is applied. The problem could be briefly expressed as following.

The water mass conservation could be represented as (i.e. continuity equation):

$$\frac{\partial h}{\partial t} = \nabla h U \tag{3.37}$$

and the sediment mass conservation is then:

$$\frac{\partial C}{\partial t} = \nabla \bullet (D \nabla C) - \nabla \bullet (UC) \tag{3.38}$$

where U, C are the depth-average velocity and concentration.

In discretized form, they are:

$$V_i^{t+\Delta t} - V_i^t = \Delta t (\sum Q_{in} - \sum Q_{out}) \tag{3.39}$$

and the sediment mass conservation is then:

$$V_i C_i^{t+\Delta t} - V_i C_i^t = \Delta t (\sum Q_{in} C_{in} - \sum Q_{out} C_{out}) \tag{3.40}$$

where Q, V are the inflow/outflow fluxes and volume.

During a computational time step Δt, when the sediment transport gradient is not balanced, the bed level is changed. So when erosion happens, the control volumes of water increase a bit, while sediment happens, the control volumes of water get decreased.

There are several possible solutions for this problem. One way is to modify the water level after the bed level changes. The original water volume is thus maintained. The change of water level might introduce certain surface wave, which would be diffused by the hydrodynamic models. The underlying assumptions are that the sediment source from the bed is infinite and the volume of sediment particles in water column is negligible. Li et al. (2010) estimated the water level perturbance using linear stability analysis and found that generally this small perturbance would be damped crossing a few grid cells within a few time steps, even though the nonlinear stability is still under exploration.

Another way is to correct the transport of water and sediment in each time step. The total transport capacity could carry a certain amount of water and sediment particles. If under erosion, more sediment particles are transported, then less water is transported. However, the underlying assumption is that the porosity of eroded sediment and deposited sediment is similar, which probably not true in general. And when the morphological factor is applied, the exchange mass between water and bed is distorted, which also increases the difficulties for the mass correction.

As a compromise, Postma (Leo Postma, personal communication, 2009, 2010) proposed that for water and most passive substance (salinity and temperature, etc.), the mass conservation is assured. For sediment, if the option one is applied, a small jump of sediment concentration is expected with the water level correction. Or if the option two is applied, the sediment concentration is a smooth curve against time while mass conservation is violated. In the end of the morphological step, all mass across the boundaries of the control volume could be summed up and be removed. This "net removal" + "morphological factor" × ("net outflow across boundaries" + "net increase in water concentration") must be exactly equal the sediment erosion/deposition of the bed during the morphological step.

3.5 Velocity integration algorithm

Velocity integration algorithm is to define a unified approach to compute the representative velocity vector in cell center. It is a critical step to assure the model generic for both structured grid and unstructured grid models.

First of all, based on the flux in/out of the volume, the representative velocity vector is integrated to the grid cell (segment) centre independent of the dimension of the grid cell.

For the bed level updating, the sediment transport flux is computed with the integrated represented velocity at the grid cell (segment) centre. It is also independent to the dimension of the grid cell.

The output of the flow data is written to a communication file, and processed by the couple program, which is presently embedded in the Delft3D-FLOW model, again not necessarily. If the other flow model, e.g., TELEMAC, etc. can generate the communication files of the same format, the morphological module is ready to be applied.

3.5.1 Generic velocity integration algorithm

Staggered finite volume based discretization are used in Delft software system. As a result, the velocities are known at the cell interfaces and even there only the component perpendicular to the cell interface is known. For sediment transport and many other practical applications, the velocity magnitude and optionally the velocity direction must be given at the cell centers where most other quantities (water level, salinity, concentrations, and other coefficients etc.) are known. The following algorithm is developed to obtain the velocity vector from the perpendicular components at the grid cell interfaces only.

Figure 3.3 shows a curvilinear grid cell used currently in present version of Delft3D system. To compute the track of a particle through the grid cell shown, one linearly interpolates the staggered velocity components independently. The normal velocities are assumed to be constant along the faces while a linear transition is assumed from

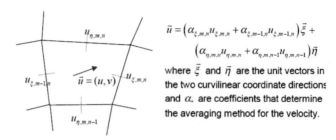

$$\bar{u} = \left(\alpha_{\xi,m,n} u_{\xi,m,n} + \alpha_{\xi,m-1,n} u_{\xi,m-1,n} \right) \bar{\xi} +$$
$$\left(\alpha_{\eta,m,n} u_{\eta,m,n} + \alpha_{\eta,m,n-1} u_{\eta,m,n-1} \right) \bar{\eta}$$

where $\bar{\xi}$ and $\bar{\eta}$ are the unit vectors in the two curvilinear coordinate directions and α_* are coefficients that determine the averaging method for the velocity.

Figure 3.3: Curvilinear grid cell with definition of velocity components

one interface to the opposing one. For the purpose of tracking particles the velocity component parallel to the interface where $u_{\xi,m,n}$ is computed is thus given by a linear combination of the velocities $u_{\eta,m,n}$ and $u_{\eta,m,n-1}$; this leads to a discontinuity in the parallel velocity component from one grid cell to the next. For other purposes the same component may be obtained from averaging the four surrounding velocity components $u_{\eta,m,n}$, $u_{\eta,m,n-1}$, $u_{\eta,m+1,n}$ and $u_{\eta,m+1,n-1}$. Figure 3.4 shows a generic volume with a couple of interfaces with neighboring cells.

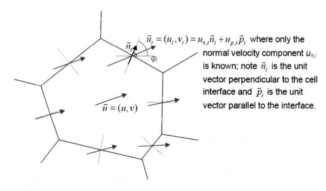

$\bar{u}_i = (u_i, v_i) = u_{n,i}\bar{n}_i + u_{p,i}\bar{p}_i$ where only the normal velocity component $u_{n,i}$ is known; note $\bar{n}_i$ is the unit vector perpendicular to the cell interface and $\bar{p}_i$ is the unit vector parallel to the interface.

Figure 3.4: Generic volume with definition of velocity components

The starting premise is that the velocity vector at the cell centre is given by some linear combination of the velocity vectors at the cell interfaces:

$$\bar{u} = \sum_i \alpha_i \bar{u}_i \Big/ \sum_i \alpha_i \qquad (3.41)$$

The linear combination, i.e. what factors α_i to use, is discussed later in this part. For the time being a generic linear combination is assumed. Equation 3.41 might be rewritten as

$$\vec{u} \sum_i \alpha_i = \sum_i \alpha_i (u_{n,i} \vec{n}_i + u_{p,i} \vec{p}_i) \tag{3.42}$$

where the parallel velocity component u_p, i are unknown. To solve this equation, the formula $u_{p,i}$ has to be specified. The most simple approximation for up, i is to assume that it equal to the component of the velocity vector in the cell centre that is parallel to the considered interface

$$u_{p,i} = \vec{u} \cdot \vec{p}_i \tag{3.43}$$

Note that this formulation has two effects: Firstly, the number of unknowns in equation 3.42 is reduced to 2 which make the vector equation solvable, and secondly, a discontinuity occurs in the parallel velocity at cell interfaces since the two neighboring cells will use different parallel components (namely, the parallel components derived from their own velocity at cell centre). The latter effect might be counterintuitive but it is consistent with the particle tracking case on the curvilinear grid as discussed above. After substitution equation 3.42 becomes

$$\sum_i \alpha_i = \sum_i \alpha_i [u_{n,i} \vec{n}_i + (\vec{u} \cdot \vec{p}_i) \vec{p}_i]$$
$$= \sum_i \alpha_i u_{n,i} \vec{n}_i + \sum_i \alpha_i (\vec{u} \cdot \vec{p}_i) \vec{p}_i \tag{3.44}$$

or equivalently (collecting terms containing at the right hand side)

$$\sum_i \alpha_i u_{n,i} \vec{n}_i = \sum_i \alpha_i \vec{u} - \sum_i \alpha_i (\vec{u} \cdot \vec{p}_i) \vec{p}_i$$
$$= \sum_i \alpha_i [\vec{u} - (\vec{u} \cdot \vec{p}_i) \vec{p}_i]$$
$$= \sum_i \alpha_i (\vec{u} \cdot \vec{n}_i) \vec{n}_i \tag{3.45}$$

where the last step results from the fact that $\vec{n}_i$ and $\vec{p}_i$ form a local normal coordinate system. The right hand side of this equation can be expanded to

$$\sum_i \alpha_i u_{n,i} \vec{n}_i = \sum_i \alpha_i [(u\vec{x} + v\vec{y}) \cdot \vec{n}_i] \vec{n}_i$$
$$= \sum_i \alpha_i [(u(\vec{x} \cdot \vec{n}_i) + v(\vec{y} \cdot \vec{n}_i)] \vec{n}_i \tag{3.46}$$

This vector equation can be rewritten as a system of equations for the two coordinate directions

$$\sum_i \alpha_i u_{n,i}(\vec{n}_i \cdot \vec{x}) = \sum_i \alpha_i [(u(\vec{x} \cdot \vec{n}_i) + v(\vec{y} \cdot \vec{n}_i)]\vec{n}_i \cdot \vec{x}$$

$$= \sum_i \alpha_i u(\vec{n}_i \cdot \vec{x})^2 + \sum_i \alpha_i v(\vec{n}_i \cdot \vec{y})(\vec{n}_i \cdot \vec{x})$$

$$= u \sum_i \alpha_i (\vec{n}_i \cdot \vec{x})^2 + v \sum_i \alpha_i (\vec{n}_i \cdot \vec{y})(\vec{n}_i \cdot \vec{x}) \qquad (3.47)$$

$$\sum_i \alpha_i u_{n,i}(\vec{n}_i \cdot \vec{y}) = u \sum_i \alpha_i (\vec{n}_i \cdot \vec{y})(\vec{n}_i \cdot \vec{x}) + v \sum_i \alpha_i (\vec{n}_i \cdot \vec{y})^2 \qquad (3.48)$$

Together the equations 3.47 and 3.48 form a system of two equations with two unknowns (u and v) with fairly complex coefficients depending on the directions of the various interfaces. The solution to this system of equations reads

$$u = \frac{1}{\Omega} \left[\sum_i \alpha_i (\vec{n}_i \cdot \vec{y})^2 \sum_j \alpha_j u_{n,j}(\vec{n}_j \cdot \vec{x}) - \sum_i \alpha_i (\vec{n}_i \cdot \vec{y})(\vec{n}_i \cdot \vec{x}) \sum_j \alpha_j u_{n,j}(\vec{n}_j \cdot \vec{y}) \right] \quad (3.49)$$

$$v = \frac{1}{\Omega} \left[-\sum_i \alpha_i (\vec{n}_i \cdot \vec{y})(\vec{n}_i \cdot \vec{x}) \sum_j \alpha_j u_{n,j}(\vec{n}_j \cdot \vec{x}) + \sum_i \alpha_i (\vec{n}_i \cdot \vec{x})^2 \sum_j \alpha_j u_{n,j}(\vec{n}_j \cdot \vec{y}) \right]$$

$$(3.50)$$

where:

$$\Omega = \sum_i \alpha_i (\vec{n}_i \cdot \vec{x})^2 \sum_j \alpha_j (\vec{n}_j \cdot \vec{y})^2 - \left[\sum_i \alpha_i (\vec{n}_i \cdot \vec{y})(\vec{n}_i \cdot \vec{x}) \right]^2 \qquad (3.51)$$

and given the expression:

$$\vec{n}_i \cdot \vec{y} = sin(\phi_i) \qquad (3.52)$$
$$\vec{n}_i \cdot \vec{x} = cos(\phi_i) \qquad (3.53)$$

Eq. 3.49 and Eq. 3.49 turn to be:

$$u = \frac{1}{\Omega} \left[\sum_i \alpha_i sin^2(\phi_i) \sum_j \alpha_j u_{n,j} cos(\phi_j) - \sum_i \alpha_i sin(\phi_i) cos(\phi_i) \sum_j \alpha_j u_{n,j} sin(\phi_j) \right]$$

(3.54)

$$v = \frac{1}{\Omega} \left[- \sum_i \alpha_i sin(\phi_i) cos(\phi_i) \sum_j \alpha_j u_{n,j} cos(\phi_j) + \sum_i \alpha_i cos^2(\phi_i) \sum_j \alpha_j u_{n,j} sin(\phi_j) \right]$$

(3.55)

where:

$$\Omega = \sum_i \alpha_i cos^2(\phi_i) \sum_j \alpha_j sin^2(\phi_j) - \left[\sum_i \alpha_i sin(\phi_i) cos(\phi_i) \right]^2$$ (3.56)

3.5.2 Discussion

Hereby three options for average factor α_i are discussed:

1. $\alpha_i = 1$;
   It would be just an arithmetic mean of the individual velocities not taking into account the fact that some interfaces may be longer than others. There is no continuum analogue.

2. $\alpha_i = B_i$;
   B_i is the width of the interface, would be an averaged based on interface widths. In this case the continuum equivalent of Eq. 3.41 reads:

   $$\vec{u} = \frac{1}{L} \oint \vec{u}(x, y) dl$$ (3.57)

   and L refers to both the combination of all interfaces and the overall length of the interfaces.

3. $\alpha_i = H_i B_i$;
   H_i is the water depth of the interface, would be an averaged based on unit dis-

charges. In this case the continuum equivalent of Eq. 3.41 reads

$$\vec{u} = \oint \vec{u}(x,y)h(x,y)dl / \oint h(x,y)dl \qquad (3.58)$$

In the current online morphology module of Delft3D-FLOW, the following expression is used:

$$U = \frac{U_{m,n}H_{U,m,n} + U_{m-1,n}H_{U,m-1,n}}{H_{\zeta,m,n}} \qquad (3.59)$$

where:

    U: refers to the velocity in ξ or m direction;

    H_U: water depth at velocity point (cell interface);

    H_ζ: water depth at water level point (cell center).

Note that the width of the cell interfaces is not taken into account (assuming homogeneity of the grid) and note that if $H_\zeta < H_U$ you may get an increase in velocity via this interpolation method (an alternative implementation where the $H_{\zeta,m,n}$ is replaced by $max(H_{U,m,n}, H_{U,m-1,n})$ was shown to be more stable during drying and flooding). Furthermore, the curvilinearity of the grid is neglected via the assumption of orthogonality.

In this study, method 3 is used for α_i because it is similar to the current implementation in the Delft3D online morphology module in structured grid and it is valid for unstructured grid as well. Furthermore, this method makes most sense for the following cases (Figure 3.5). Given a stationary flow in a narrowing channel the following behavior is observed, constant discharge Q and cross sectional area $A >> a$ and thus $u << U$. Using method 1 one gets as average velocity $(U + u)/2 \approx U/2$. In method 3 one gets as average velocity $(Q + Q)/(A + a) \approx 2Q/A = 2u$. The latter result gives a lower velocity more in line with what could be expected based on linear interpolation. Method 2 is not well grounded as well since this formula takes into account the different lengths of the interfaces but neglects the variation of the water depth, which might be valid in deep sea but is not appropriate in shallow water areas.

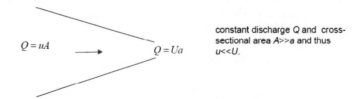

Figure 3.5: Generic effect of the cross-sectional area on the average velocity

Furthermore, the analysis above does not work for 1D model since the velocity in 1D models is only a scalar quantity and not a vector quantity.

3.6 Model Structures and dataflow

3.6.1 Structures and Dataflow

As described above, the generic geomorphological model consists of six modules, namely the hydrodynamic module, including current module (and turbulence closure module), wave module, used to provide the hydrodynamic information, the bed state description module to describe the bed composition, the sediment transport module to provide the sediment transport flux, the geomorphologic bed level update module to update the bathymetry and to give the feedback to other five modules, and the vegetation population dynamic module to provide the instant spatial pattern and vegetation properties, which will be discussed in Chapter 5. The implementation chart demonstrates the six modules in the model (Figure 1.1). The later four modules are extended or developed using the Open Process Library (OPL) in the frame of Delft3D-WAQ.

Open processes library is a collection of hundreds of substances, corresponding processes and parameter sets. The library could be accessed by a GUI configuration tool, namely PLCT (processes library configuration tool).

From the modeler point of view (Figure 3.6), modeler may define the substances, processes firstly, then define the input file (contains grid, time information, etc.). Thus the Delft3D-WAQ preprocessor would sort out the substances and linkages between the processes. In the end the Delft3D-WAQ processor could solve the equation systems by more than 20 types of numerical solvers. From the developer point of view (Figure 3.6), the third party functions could be compiled into an extra library and would be called by the Delft3D-WAQ processor.

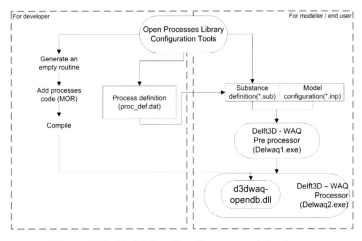

Figure 3.6: Model flowchart for users and developers

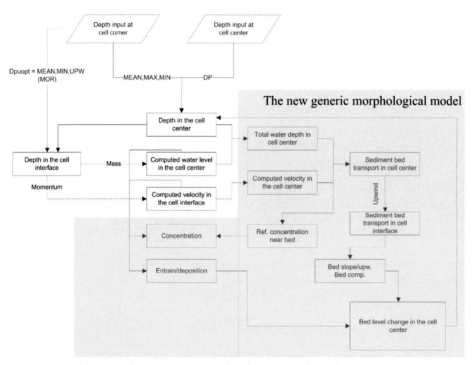

Figure 3.7: Data flow diagram: Water depth. The yellow block indicates the generic morphological model.

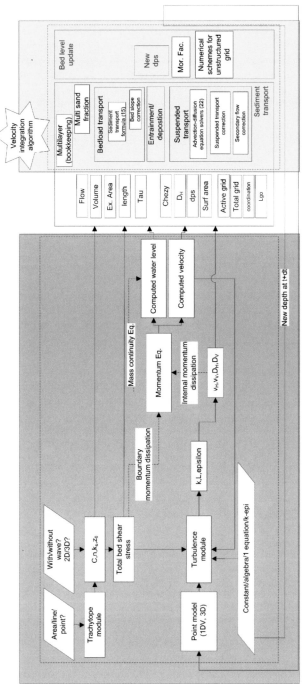

Figure 3.8: Data flow diagram: Bed roughness (refer to Figure 5.3). The yellow block indicates the generic morphological model.

One of the most important variables in morphological module is the water depth. The dataflow for a discritisized water depth is described as the following (Figure 3.7). The depth includes sum of two parts, the water depth, varying with the morphological change and the water level, which is modified for every flow simulation time step. Another important variable in hydrodynamic and morphological module is the bed roughness coefficient. The dataflow on the bed roughness processing is described in Figure 3.8. The bed roughness coefficient quantifies the friction of the flow, determines the conveyance capacity and thus entrains sediment particles. One of the approaches to take into account the effect of vegetation also is formulated related to the change of the bed roughness coefficient (refer to Section 5.3.1, Section 5.4, and Figure 5.3).

The results from this new generic morphological model are compared with that from Delft3D online MOR version, and results from analytical solutions and flume experiments.

3.6.2 Substances and processes definition

In the new morphological model, four sediment fractions can be defined as the transportable or non-transportable substances at this stage. The number of sediment fraction would be able to extend as many as 99 sediment fractions in near future. Each sediment fraction has two corresponding parts. One part is treated as bed load and the other is treated as suspended load. The available 24 physical processes could be classified in three groups. The first group of processes is used by each sediment fraction. The second group is used by all the fractions in the model. The third group includes the most parameter calculation, which is to be called by all the fractions as well. The detail description of the function, process ID, process name, input and output of each processes are listed in Figure A.2, A.3, and Figure A.1.

The initialize process (The 7th process in Figure A.3) starts up the bed book keeping system, and keeps the maintenance.

Processes for all the sediment fractions showed in Figure A.3 is used to prepare the crucial coefficients, such as, D_{10}, D_{35}, D_{50}, D_{90} and D_{mean}, the properties (volume, exchange area, water depth, velocity vector at cell center, wave parameters) of the cell, vertical profile of diffusivity coefficient, and so on.

The corresponding processes listed in Figure A.1 provide general parameters, such as, fluid density, kinetic viscosity, which is a function of ambient salinity and temperature, bed roughness prediction, bed form prediction, settling velocity based the results from the processes mentioned above, and critical shear stress, critical velocity etc.

Processes attached for each sediment fraction includes (Figure A.2):

- Sediment diameter (Process A in Figure A.2)
  In this model, most flexibility for the sediment diameter input is offered. Either the

D_{50} directly given, or a range of sediment distribution curve could be input and this process will compute (or estimate) parameters, such as, sediment diameter, volume fraction, mass fraction etc..

- Deposition (Process B in Figure A.2)
  This process is responsible to exchange the sediment in the water column and the bed. A sink term is stored for the transport solver, while the term acts as a source term for the bed book keeping system for each fraction.

- Resuspension (Process C in Figure A.2)
  The same as the deposition processes. This process is responsible to exchange the sediment in the water column and the bed. A source term is stored for the transport solver, while the term acts as a sink term for the bed book keeping system for each fraction.

- Hiding/exposure correction (Process D in Figure A.2)
  The source/sink term from the deposition/resuspension processes is corrected by the hiding and exposure coefficient, following Egiazaroff and other three formulations (Deltares, 2010a).

- Settling velocity and correction (Process H in Figure A.1)
  This process provides settling velocity, which is corrected by hinder settling processes if the sediment fraction is mud.

- Sediment transport capacity at cell center (Process E in Figure A.1)
  More than 15 sediment transport formula are programmed to compute the sediment transport capacity. The formula are described in Deltares (2010a).

- Sediment transport capacity at exchange (Process F in Figure A.1)
  The sediment transport capacity at cell center is assigned to the exchange using the numerical schemes described in Section 3.4.3.

- Sediment transport flux for whole cell (Process G in Figure A.1)
  Volumetric sediment transport flux is computed based the sediment transport capacity at exchange.

- Bedslope correction
  The sediment transport flux then is corrected by bed slope effect. Refer to Section 3.3.

- Secondary flow correction
  For 2D simulation, the vertical spiral motion is simulated by a transport (advection-diffusion) equation. This process is optional in this model.

The final process AVOL (the 9th process in Figure A.3) is to compute the volume for the water column. Following the finite volume approach, the sediment transport fluxes are integrated along the exchanges and bottom boundary, and the deposition/erosion of the bed level at corresponding grid cell (segment) is then determined.

3.6.3 Linking the modules

The communication between various modules, i.e. between the different computer processes at low level, takes place either through communication files, which can be binary or ASCII files, or through a named-pipe in memory (i.e., DelftIO). Both optional methods are in this model.

The flow module outputs the results in group as binary communication files, and the coupling program reads the communication files and changes them into a set of files in Delft3D-WAQ format. Delft3D-WAQ passes the flow information to the morphological module. After morphological module is activated, the bed level changes are calculated and passed back to the flow module by a named pipe through memory. In the next time step, the flow module computes the flow information based on the updated bed level, and the same procedures loop till the end of the simulation time. Notice that when the information is transferred between computer processes through the named pipe, the computer processes are not necessarily running on the local computers. They could be even remote computers connecting by internet.

3.7 Conclusions

A generic geomorphological model is built in this chapter. by extending some existing modules and developing new modules. The model is generic in two senses: One is that this geomorphological model functionally can be coupled with different hydrodynamic models using structured and unstructured grid, for FVM, FEM and FDM hydrodynamic models, such as Delft3D-FLOW based on curvilinear structured grid, D-Flow FM based on a combination of 1D, 2D and 3D unstructured grid, TELEMAC based on triangular unstructured grid and SOBEK based on a river network schematization. Other hydrodynamic models, such as ROMS, ADCIRC ect. could use this morphological model directly if they could provide the input files in format of Delft3D-WAQ. On the other hand, the generic model is supposed to be robust and cover a wide range of processes at various scales.

The genericity of this geomorphology model is assured by the adaptive numerical algorithms and the open model structure. The numerical algorithms include the bed updating algorithm for both structured and unstructured grid and the generic velocity integration algorithm.

The advantage of this model includes:

- Flexibility
  The morphological model is independent to flow model. It can be coupled with many

other hydrodynamics model tools, such as: 1D (SOBEK, by river group), 2D (Delft3D-FLOW), 3D (Delft3D-FLOW) flow model, no matter the flow model is integrated with FDM, FEM, FVM, discretized into triangle grid, curvilinear grid, pentagon, etc.

- Extendibility
  It is easy to extend. 4 modules are extended/developed in the open frame of Delft3D-WAQ, i.e. the bed state description module (book keeping the bed composition), the sediment transport module (suspended load modeled by solving transport equations / source and sink term, bed load modeled by numbers of empirical formula, dumping, dredging), bed level updating module. And the open structure accompanying with ecological processes, it is a good tool for bio-geomorphology research.

- Wide-usability
  It inherits the advantages of Delft3D-WAQ system, namely, more than six hundred chemical / bio / ecological processes have been validated and are ready to use; the mass conservation is guaranteed by definition.

In Chapter 5, ecological processes for aquatic vegetations will be included in this generic geomorphological model. And in future studies, wave influence and aeolian transport might be included as well. Thus this generic geomorphological model is not only a morphological model by then, but it becomes a generic platform for interdisciplinary studies potentially.

In the following Chapter 4, validations for this generic geomorphological model are carried out. The validation cases cover a wide range of morphological processes at different scales.

Chapter 4

Validation of the Generic Coastal

Morphological Model

4.1   Introduction

Validation of a new numerical model can be subdivided into two parts: validation of process formulations and validation of code for a range of geometries. Since the flexible mesh flow solver is not yet available, we cannot test the full intended scope of the geometries. Hence, this chapter focuses on the validation of process formulations only using the hydrodynamic conditions from Delft3D FLOW curvilinear models. All the sediment transport and morphological process formulations are newly developed based on unstructured grid methods described in Chapter 3.

One of the advantages of the process-based numerical model is that the model can represent processes over quite a wide range of scale (Lesser, 2009). Unlike physical scale models, numerical models can represent the processes at real scale in case the processes are resolved in the numerical model system. The validation of this generic geomorphological model includes five test cases. These cases are generally used as for validations of various morphological models (e.g. Lesser et al. (2004),Callaghan et al. (2006) and Villaret and Gonzales (2005)).

The test cases are compared to existing analytical solutions, flume experiments and other verified numerical simulation results, such as, from Delft3D-FLOW online MOR and Telemac. Sediment transport and morphology processes are verified separately in the testcases. Case 1 highlights the bedload transport and bedslope correction at small scale (flume scale, 2D) (Section 4.2.1). Effects of different numerical schemes discussed in Section 3.4.3 are shown as well. Case 2 highlights the bedload transport at large scale (river reach scale) (Section 4.2.2). Case 3 demonstrates the development of longitudinal and vertical profiles of suspended sediment transport (Section 4.2.3). Case 4 shows bedload transport, suspended sediment transport, morphological update in 2D groyne fields (Section 4.2.4). Case 5 is verified against the trench migration experiment in a flume carried out by Van Rijn in 1987 (Section 4.3.1). Bedload transport, bedslope correction, suspended sediment transport, suspended sediment concentration profile in flume scale are compared with measurements and results from Delft3D-FLOW online MOR. Furthermore, all the validation cases in this chapter are carried out with the newly developed generic geomorphological model, which is theoretically validated using unstructured grid. However, since the flexible mesh flow solver is not yet available, we cannot test the full unstructured grid and Delft3D-FLOW is applied to provide the hydrodynamic information. Hereafter if the velocity integration approach (Equation 3.59) is used, we call the new generic geomorphological model Structured Generic MOR. If the velocity integration approach (Equation 3.58) is used, we call it Unstructured Generic MOR.

4.2 Validation against analytical solutions

4.2.1 Hump migration (2D/3D, flume scale)

This test case is to validate the convection process of hump migration in a channel, which can be compared with analytical solution and other morphological models, such as Delft3D online MOR and Telemac SISYPHE. A series of tests are performed simulating bedload transport process, suspended sediment transport with and without bedslope correction. The different numerical scheme effects are also checked.

The test case is set up based on a 16 meters long, 1 meter wide straight flume with constant water depth of 60 cm. The imposed average flow velocity in the flume is 0.5 m/s. The initial bottom is horizontally flat except in the middle cross section of the flume, 2m from the inflow boundary a 10 cm high sand hump is constructed. The initial shape of the hump is described as the following function:

$$z_{hump} = 0.1 sin^2 \left(\frac{\pi(x-2)}{8} \right) \qquad\qquad x \in [2, 10]$$

$$= 0 \qquad\qquad\qquad otherwise \qquad\qquad (4.1)$$

Given this condition, the Froude number in this case would be around 0.16 - 0.22. Based on Bernoulli equation, the subsequence water level difference caused by the hump is in the magnitude of ~1cm, thus the non-hydrostatic effect should be negligible. Initial bed material is uniform sand with d_{50} of 0.141 mm. Sediment density $\rho_s = 2650 \ kg/m^3$, and the coefficient of the porosity is 0.4. The friction coefficient (*Chézy*) C $= 45.57 \ m^{1/2}/s$ ($k_s = 0.021$ m).

Numerically the domain is discretized into 80 × 20 grid cells with size of 20 cm × 10 cm. Time step used here is 1 second and simulation results of 5 morphological hours are compared with quasi-analytical solutions.

With assumption of rigid bottom and rigid lid condition at water surface, it is assumed that there is no variation at the free water surface. In this simple case (1D situation, no suspension), the conservation equation simply writes (refer to Eq. 3.11):

$$(1-p) * \frac{\partial z_b}{\partial t} + \frac{\partial q_{s,x}}{\partial x} = 0 \qquad\qquad (4.2)$$

where:
$q_{s,x}$ here represents sediment transport flux (volumic with unit m^2/s); z_b is the bed level (m); p is the porosity.

A typical bedload transport can be written as:

$$q_{s,x} = a\bar{u}^b = a \left(\frac{Q}{h} \right)^b \qquad\qquad (4.3)$$

so, Eq. 4.2 can be easily transform as a following advection equation:

$$\frac{\partial h}{\partial t} + c \frac{\partial h}{\partial x} = 0 \qquad\qquad (4.4)$$

where the celerity of the sediment on the solid / liquid interface is $c = \frac{1}{1-p} b \frac{q_{s,x}}{h}$.

If the Engelund and Hansen formula is used:

$$q_{s,x} = \frac{0.05}{(s-1)^2 g^{0.5} d_{50} C^3} \tag{4.5}$$

with b = 5, the celerity and displacement of the hump can be easily computed. The advection equation is solved numerically by the method of characteristics and the distance of displacement of a sand hump along the x axis (convection) is obtained by multiplying the celerity by the time-step.

Figure 4.1 shows the results from the numerical modeling and the analytical solution. The general behavior of the numerical modeling results (displacement and celerity) shows that the transport processes are accurately represented by the new Generic MOR model.

Figure 4.1a shows the streamwise bed level profiles computed by Delft3D-FLOW online MOR and the Structured Generic MOR. In the Structured Generic MOR, the flow velocity is integrated to the cell center similar to the present algorithm in Delft3D-FLOW online MOR with the assumptions: 1) the width of the cell interfaces is not taken into account; 2) the curvilinearity of the grid is neglected via the assumption of orthogonality (refer to Section 3.5.2). Figure 4.1b shows the difference by applying different velocity integration algorithm: with or without the water depth of the interface, averaged based on unit discharges (Eq. 3.58 in Section 3.5.2). Because of the regularity of the grids in this particular case, consistency of the two integration methods is expected and is also presented in the results. In the present version of Delft3D-FLOW online MOR, 1st-order upwinding method is applied for its stability property. As expected, numerical damping of the hump peak and smaller bed level slope in the hump tail area are also found in Figure 4.1a, b. Sediment transport flux computed by Van Rijn 93 formula is comparable to the results from EH formula. Figure 4.1c shows that the hump displacement, hump peak damping and the celerity are similar to those shown in Figure 4.1a, b, e. Figure 4.1d shows that streamwise bed level profiles computed by Delft3D-FLOW online MOR and the new Generic MOR with bedslope correction (Bagnold, Ikeda and Van Rijn algorithm). Results computed by the new Generic MOR are reasonably well reproduced compared to Delft3D-FLOW online MOR. Figure 4.1e shows that with central method, the results from both FLOW online MOR and the new Generic MOR model are accurate, but the results are ruined by the artificial wiggles after 4 hours.

From Figure 4.1a to e and Figure 4.1f, the results computed by both Telemac SISYPHE and Delft3D system are well comparable with the analytical solution. The model Telemac SISYPHE gives out quite good accuracy and similar accuracy is given by the new Generic MOR using central method (Figure 4.1e). A two step approach is used in SISYPHE. Firstly a prediction of sediment transport flux is made followed by a correction step by θ averaging to smooth out the wiggles to some extend, which makes the SISYPHE accurate during the first few hours. However, if the simulation goes longer, spurious wiggles are bound to occur as well.

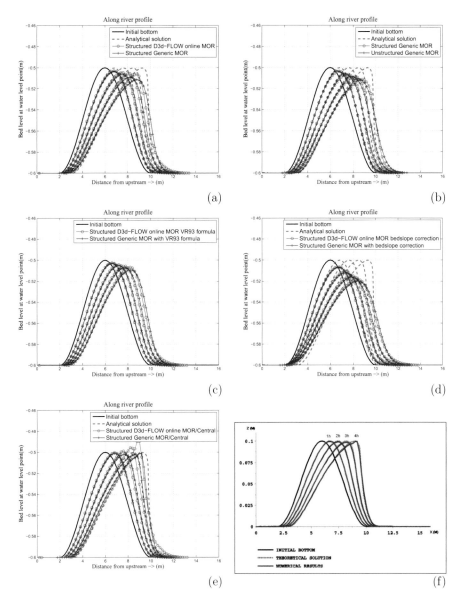

Figure 4.1: Hump migration along the flume in 5 hours. Results from Delft3D-FLOW online MOR, the generic MOR and analytical solution. (a) Structured Generic MOR vs. Delft3D-FLOW online MOR (upwinding, EH formula); (b) Unstructured Generic MOR vs. Unstructured Generic MOR (upwinding, EH formula); (c) Structured Generic MOR vs. Delft3D-FLOW online MOR, VR93 formula (upwinding); (d) Structured Generic MOR vs. Delft3D-FLOW online MOR, with slope correction (upwinding, EH formula, alfabs = 1.0, alfabn = 1.5); (e) Structured Generic MOR vs. Delft3D-FLOW online MOR, central method, no slope correction (EH formula); (f) Results from Telemac SISYPHE with similar model settings (EH formula).

In conclusion, the hump convection process due to bedload (total load) transport is well represented in both the new Generic MOR and Delft3D-FLOW online MOR. The new velocity integration method applied in the new Generic MOR model produces reasonable good and consistent results as other modeling tools and analytical solutions. Upwinding scheme is recommended due to its robustness and simplicity to implement. Centered scheme for bed level updating and bedslope corrections are optional and tested in the case. However, the centered scheme should be applied with special care with proper damping or a pragmatic bedslope correction. Different scenarios show that the bedload transport and morphology processes are well represented in the new Generic MOR.

4.2.2   Equilibrium slope development (2D/3D, river reach scale)

This test case is included to validate sediment erosion and convection processes of bed load transport, which can be compared with analytical solution and result from Delft3D online MOR. Again, a series tests are performed simulating bedload transport process, suspended sediment transport with and without bedslope correction.

The testing case is setup based on a long (10 kilometers) and straight river reach with width of 200 meters. Initially water depth is constant 5m. The imposed average flow velocity in the flume is around 1 m/s. The initial bottom is horizontally flat. Given this condition, the Froude number in this case would be around 0.15. Initial bed material is uniform sand with d_{50} of 0.141mm. The initial thickness of erodible bed material is uniformly 10m. Sediment density $\rho_s = 2650 \ kg/m^3$, and the coefficient of the porosity is 0.4. The friction coefficient (*Chézy*) C = 65 $m^{1/2}/s$ (i.e., $k_s = 0.015$m).

Numerically the domain is discretized into 20 × 2 grid cells with size of 500 m × 100 m. Time step used here is 1 minute and simulation results of 3 morphological years are compared with empirical analytical solutions.

Given the *Chézy* coef of 50 $m^{1/2}/s$ and average velocity of 1 m/s, the friction velocity is around 0.048 m/s, which gives out the equilibrium slope of 5×10^{-5}. With a 10 km long river reach, the bedlevel difference is approximately 0.5 m.

Initially, the river bed is flat. The model starts with a hot start. The upstream boundary is water level constant, and the downstream is constant discharge. The energy slope is much higher than the water surface slope. Thus accelerating flow is generated, which causes the increasing sediment transport rate along the river, and erosion. The equilibrium slope is developed and the bed slope is adapted with the energy slope. Because in the FLOW online MOR, the morphology point before the computation cell is fixed, while the new Generic MOR is fixed in the first cell, the bed level from these 2 models are 2 lines in parallel with each other (Figure 4.2). However, the water levels from 2 models are equally good compared with the analytical approximations (Figure 4.3). In conclusion, the sediment transport and morphology processes in river reach scale are well represented in the new Generic MOR model.

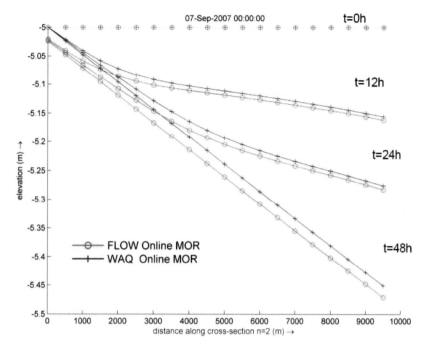

Figure 4.2: Bed level at water level point, longitude profile. Red line with circle is from Delft3D-FLOW online MOR and blue line with plus is from the new Structured Generic MOR.

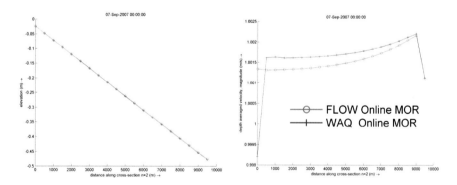

Figure 4.3: Water level longitude profile (left panel) and depth averaged velocity magnitude (right panel)Red line with circle is from Delft3D-FLOW online MOR and blue line with plus is from the new Structured Generic MOR.

81

4.2.3   Suspended sediment transport and development of equilibrium conditions (2D/3D)

This case is to validate the sediment erosion/deposition and advection/diffusion processes of suspended sediment transport. Van Rijn 1993 and 2004 formula are tested Van Rijn (1993); Van Rijn et al. (2004). Vertical profiles of sediment concentration are compared with result from Delft3D-FLOW online MOR.

The test case is set up based on a 1.6 kilometer long and straight river with width of 100 m. Initially water depth is constant 10 m. And the initial river bed has equilibrium slope ($i = 5 \times 10^{-5}$). The imposed average flow velocity in the flume is around 1 m/s. Given this condition, the Froude number in this case would be around 0.1. Initial bed material is uniform sand with d_{50} of 0.2 mm ($\omega_s = 2.35 cm/s$). The initial thickness of erodible bed material is uniformly 10m. Sediment density $\rho_s = 2650 \ kg/m^3$, and the coefficient of the porosity is 0.4. The friction coefficient (*Chézy*) C = 50 $m^{1/2}/s$ ($k_s = 0.088$m).

Numerically the domain is discretized into 80×10 grid cells with size of 20m×10m. Time step used here is 6s and simulation results of 16 morphological hours are compared with empirical analytical solutions.

By integration of the balance settling and diffusion,

$$D_V \frac{\partial c}{\partial z} + \omega_s c = 0 \tag{4.6}$$

and assuming a parabolic eddy viscosity profile:

$$D_V = \frac{\kappa u_* h}{\sigma_T} \frac{z}{h} \left(1 - \frac{z}{h} \right) \tag{4.7}$$

where: κ is the Von Kármán constant, u_* the friction velocity and σ_T the Prandtl-Schmidt number relating eddy diffusivity and eddy viscosity. The Rouse solution is given out:

$$\frac{c}{c_a} = \left[\frac{a(1 - z/h)}{z/h(1 - a)} \right]^{\gamma} \tag{4.8}$$

where c_a is a reference concentration at level a and with Rouse number $\gamma = \frac{\sigma_T \omega_s}{\kappa u_*}$.

Furthermore, for $\gamma < 1$, if Eq. 4.8 integrated over depth gives:

$$\bar{c} = \int_{z/h=0}^{z/h=1} c \; d\left(\frac{z}{h}\right) \tag{4.9}$$

$$= c_a \left(\frac{a}{1-a}\right)^{\gamma} \int_0^1 \left(\frac{1-z/h}{z/h}\right)^{\gamma} d\left(\frac{z}{h}\right) \tag{4.10}$$

$$= c_a \left(\frac{a}{1-a}\right)^{\gamma} \frac{\pi\gamma}{sin(\pi\gamma)} \tag{4.11}$$

Thus, substituted back to Eq. 4.8, gives:

$$c = \bar{c}\frac{\pi\gamma}{sin(\pi\gamma)}\left(\frac{1-z/h}{z/h}\right)^{\gamma} \tag{4.12}$$

Then given the vertical coordinate, the vertical profile at each point is ready (in this case, $\sigma_T = 1$, $h = 10$m, $\omega_s = 0.0235$m/s, $\gamma = 0.75$, $c_a = 0.31g/l$, $a = k_s \sim 8.5$cm over water depth of ~ 10m).

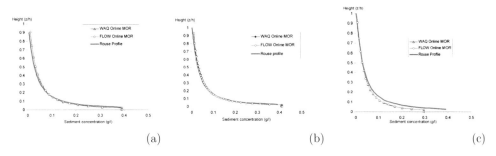

(a)   (b)   (c)

Figure 4.4: Vertical sediment concentration profile. (a) Algebraic turbulence model (20 layers in vertical); (b) Algebraic turbulence model (40 layers in vertical); (c) $k - \varepsilon$ turbulence model (20 layers in vertical). Red line with circle is from Delft3D-FLOW online MOR and blue line with plus is from the new Structured Generic MOR.

Numerical vertical sediment concentration profiles obtained from the new Generic MOR and Delft3D-FLOW online MOR agree well with the analytical Rouse profile.

Figure 4.4 shows the equilibrium vertical sediment concentration simulated with different layers numbers, turbulence closure models. Little variability is discovered. The comparisons of concentration magnitude and vertical profile show good coincidence, which implies the source/sink terms, transportation equation solvers function properly, and the implementations in the new Generic MOR model are succeed.

A more comprehensive comparison between numerical and analytical concentration profiles, which were obtained using a 1DV code, can be found in Uittenbogaard et al. (1996).

Notice that the ADI scheme (the equivalent scheme 19 in WAQ) is applied in FLOW-MOR, while the scheme 15 (an unstructured grid solver) is used in the new Generic MOR. Refer to Sec. 3.1 and Sec. 6.1 for more details.

In conclusion, the suspended sediment transport process and implementation of source/sink term are well represented in the new Generic MOR, compared to the analytical profile and Delft3D-FLOW online MOR.

4.2.4 Groyne test (2DH, river reach scale)

This case is to reproduce the morphological change adjacent to the groyne and to predict the complex horizontal morphological features. The results are compared with those from Delft3D-FLOW online MOR.

The test case is set up based on a long (20 kilometers) and straight river with width of 1 kilometer. Initially water depth is constant 5m. The imposed average flow velocity in the river is around 1 m/s. The initial bottom is horizontally flat. A 300 m groin is put at the 2 km from the upstream boundary (Figure 4.5). The width of groin is assumed to be very small. Given this condition, the Froude number in this case would be around 0.15-0.4. Initial bed material is uniform sand with d_{50} of 0.2mm. The initial thickness of erodible bed material is uniformly 10m. Sediment density $\rho_s = 2650 \ kg/m^3$, and the coefficient of the porosity is 0.4. The friction coefficient (*Chézy*) C $= 65 \ m^{1/2}/s \ (k_s =0.015\text{m})$. Numerically the domain is discretized into 200 × 10 grid cells with size of 100 m × 100

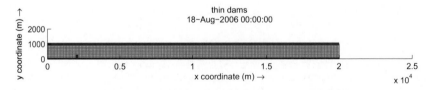

Figure 4.5: The groin and the computational grid

m. Time step used here is 60 seconds and simulation results of 400 morphological days are compared with the results from Delft3D-FLOW online MOR. The model also starts with a hot start. The results show the morphology patterns in horizontal (Figure 4.7). Both model produced similar irregularity along the channel. The morphology patterns are also well reproduced. The significance of bedslope correction is showed that the deep channel and high hump are avoided in Figure 4.6.

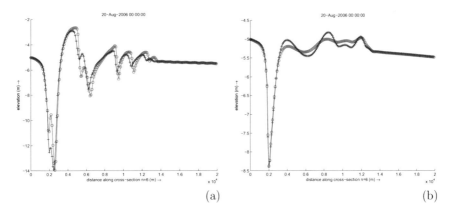

(a)   (b)

Figure 4.6: Simulated bedlevel along the middle riverline (6th gridline) after 400 morphological days. Notice that the data range of y-axis for these two figures are significantly different. (a) Without bedslope correction; (b) With bedslope correction ($\alpha_{bs} = 1.0$, $\alpha_{bn} = 20$). Red line with circle is from Delft3D-FLOW online MOR and blue line with plus is from the new Structured Generic MOR.

In conclusion, the horizontal patterns are well represented in the new Generic MOR model, as well as Delft3D-FLOW online MOR, and sediment transport under horizontal complex bed condition is verified.

4.3 Validation against physical experiments

4.3.1 Trench Migration

This case is to compare observed trench migration with migration after 15 hours in a flume experiment of van Rijn 1987. The simulation results from Delft3D-FLOW online MOR is also compared. Specifically, a) To validate the 3D advection-diffusion process of suspended transport; b) To validate bed load transport in 3D cases; c) To validate the vertical equilibrium sediment concentration profile; The trench is formed by its shape in X-direction (streamwise direction), and in Y-direction, the bathymetry is uniform. Thus this case is also a 2DV case. The advantage of this test case is the existence of a well-controlled flume experiment and also corresponding field measurement. The testing case is setup on a 30 meters' long and straight flume with width of 0.5 m. Initially water depth is constant 0.39. The imposed average flow velocity in the river is around 0.51 m/s (constant downstream discharge $0.1989 m^3/s$ approximately). The trench geometry is shown in Figure 4.8. The initial bedslope is ($i = 4 \times 10^{-4}$), while the trench depth

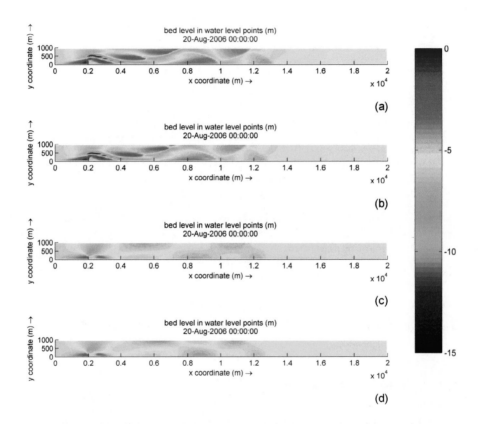

Figure 4.7: Simulated bed pattern after 400 morphological days. (a) FLOW-MOR, no bedslope correction; (b) the new Generic MOR, no bedslope correction; (c) FLOW-MOR, with bedslope correction ($\alpha_{bs} = 1.0$, $\alpha_{bn} = 20$); (d) the new Generic MOR, with bedslope correction ($\alpha_{bs} = 1.0$, $\alpha_{bn} = 20$).

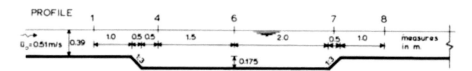

Figure 4.8: Sketch of the flume and trench geometry (after Van Rijn, 1987)

relative to the bed is 0.175 m, the slope of the trench boundary is 1:3 and the trench starts at 13.45 m measured from the inflow boundary.

Given this condition, the Froude number in this case would be around 0.25. Initial bed material is uniform sand with d_{50} of 0.16 mm and D_{90} of 0.2 mm ($\omega_s = 1.3cm/s$). The initial thickness of erodible bed material is uniformly 10m. Sediment density $\rho_s = 2650$ kg/m^3, and the coefficient of the porosity is 0.4. The friction coefficient (*Chézy*) C = 39.23 $m^{1/2}/s$ ($k_s = 0.025$ m). The total inflow sediment flux is $0.04kg/m\dot{s}$ and the bed at inflow boundary is assumed to be fixed.

Numerically the domain is discretized into 100×5 grid cells with size of 0.3 m $\times$ 0.1 m. Ten log thick layers are employed in vertical. Time step used here is 1s and simulation results of 15 morphological hours are compared with the results from Delft3D-FLOW online MOR and the measured data. The model also starts with a hot start.

As described as Lesser et al. (2004), water flows over a steep-sided trench cut in the sand bed of flume. With equilibrium sediment inflow boundary, the flow reaches the upstream of the trench with equilibrium sediment concentration profile. As the flow decelerates over the deeper trench, sediment is deposited. Then when flow accelerates over the downstream of the trench, sediment is eroded. Due to the spatial difference of sedimentation and erosion, the trench migrates. Figure 4.9 upper panel shows the sediment concentration profiles under the constant flow and sediment input conditions before the bedlevel is updated. Figure 4.9 lower panel shows the initial bedlevel and changes after 5, 10 and 15 hours. The trench is reduced to 1/2 of the initial depth and migrated for around 3 meters. Measured velocity profile and concentration profile along the flume (after Van Rijn (1987)), simulated velocity and concentration profiles from Delft3D-FLOW online MOR and the new Generic MOR model are presented. The models reproduced the experiment reasonably well. Table 4.1 gives the root mean square error of the results from Delft3D-FLOW online MOR and the new Generic MOR models. The error of bathymetry after 15 hours is less than 1 cm. The Brier Skill Score also shows the same (Table 4.2).

Using finite volume type of velocity integration algorithm (refer to Section 3.5.1) for unstructured grid, the results are also reasonable. Figure 4.10 (compared to Figure 4.9) shows the sediment concentration profiles at different positions under the constant flow and sediment input conditions before the bed level is updated. Figure 4.10 (compared to Figure 4.9) shows the initial bedlevel and changes after 5, 10 and 15 hours. It demonstrates that the finite volume type of velocity integration algorithm is accurate. Root

87

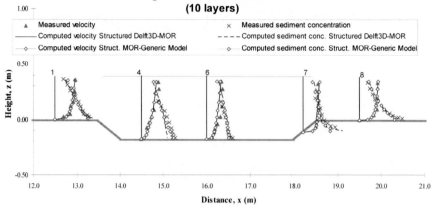

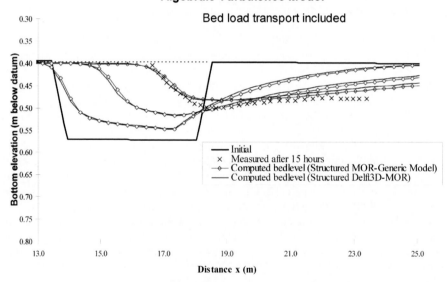

Figure 4.9: Modeling trench migration against the measurement and structured Delft3D online MOR results. Upper panel: Velocity and sediment concentration profiles across the trench along the flume before bedlevel update starts. Lower panel: Bedlevel along the flume after 5, 10 and 15 hours morphological updating. (Measurements refer to Van Rijn 1987; Delft3D-FLOW online MOR settings refer to Lesser et al, 2004)

mean square errors in Table 4.1 and the Brier Skill Score in Table 4.2 also show the same as well.

Table 4.1: RMSE for each model against measurement

Root Mean Square Error	Structured Delft3D online Mor	Structured Generic Mor	Unstructured Generic Mor
Velocity [m/s]			
Profile 1	0.0554	0.0541	0.0541
Profile 4	0.0682	0.0718	0.0718
Profile 6	0.0665	0.0668	0.0668
Profile 7	0.0356	0.0299	0.0299
Profile 8	0.0362	0.0378	0.0362
Sed. Conc. [mg/l]			
Profile 1	116.4113	77.2317	77.2317
Profile 4	285.7592	121.7987	253.5455
Profile 6	51.7146	45.9214	45.5892
Profile 7	152.9059	110.4077	76.8875
Profile 8	349.5913	659.0020	656.449
Bedlevel change [m] after 15 hours	0.0144	0.0121	0.0107

Table 4.2: Brier Skill Score of the models

	Structured Delft3D online Mor	Structured Generic Mor.	Unstructured Generic Mor.
Bedlevel change after 15 hours	0.9777	0.9843	0.9876

In conclusion, both bedload and suspended sediment transport processes and morphological update are well implemented in the new Generic MOR. The results are in very good agreement with flume measurements and Delft3D-FLOW online MOR model results.

4.4 Discussions and conclusions

The validation cases show that the new Generic MOR is able to simulate physical processes separately and in combination. The processes verified in this chapter includes: 1) sediment (sand/mud) entrainment, transport, (hindered) settling, deposition; 2) acceleration and deceleration flow; 3) bed slope effects; 4) sediment exchange with bed book keeping system and the corresponding bed roughness.

The validation results also show that the overall model structure is well designed and implemented. The communications between modules are effective. The extended sediment transport module and the new developed bed state module, the bed level updating module in the new Generic MOR are verified against the datasets from laboratory experiments and analytical solutions.

The validation cases also show that the numerical algorithms are efficient and robust as we expected. The bed level updating and sediment transport are well reproduced. Even accelerated with morphological factor, the bed level updating is quite stable. The two ways communication between modules is proven to be reliable. The model structure is demonstrated consistent. The velocity integration algorithms based on mass conservation also works fine. The velocity vectors in the exchanges of grid cell (segment) are decomposed into normal components and parallel components. Only the normal components are used and the parallel components are neglected, which implies there is still room for improvement. In addition, the bed book keeping system is operational but not properly verified due to lack of detailed data in both field and laboratory. It is found that the suspended sediment transport, thus the bed level updating, is sensitive to the bed roughness and vertical profile of diffusion coefficient, which also needs further study. Furthermore, the combination of physical processes introduces a vast amount of work in future verification (Lesser, 2009).

Nevertheless, this generic geomorphological model provides an open platform to study the combination of processes. In this chapter, physical processes are focused on. In the following chapter, ecological processes are introduced to this generic geomorphological model, thus efforts are put towards a bio-geomorphological model.

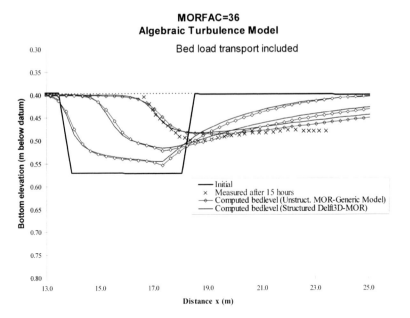

Figure 4.10: Modelled trench migration against the measurement and structured Delft3D online MOR results. Upper panel: Velocity and sediment concentration profiles across the trench along the flume before bedlevel update starts. Lower panel: Bed level along the flume after 5, 10 and 15 hours morphological updating (Measurements refer to Van Rijn 1987; Delft3D-FLOW online MOR settings refer to Lesser et al, 2004)

Chapter 5

Extension to Bio-geomorphological Modeling and Validation

5.1 Introduction

Bio-geomorphology studies the influence of plants, animals, and microorganisms on the earth surface processes and development of landforms (Viles, 1988). It is a relatively new interdiscipline that combines biology, ecology and geomorphology. Bio-geomorphology modeling is regarded as an important tool to understand and thus predict the ecological and morphological change in the system. The essence of bio-geomorphology modeling is to identify and represent the links between the hydrodynamic, morphodynamic, water quality and ecological processes involved, on the appropriate spatial and temporal scales. Over the last decade, much progress has been made. However, bio-geomorphological modeling is still a new territory in both the morphological and ecology discipline due to the large gaps between field data and models, and between the various disciplines.

This chapter aims to explore the possibilities to integrate the ecological processes into the generic geomorphological model frame described in Chapter 3. Thus firstly, efforts have been put towards to identify the scales of relevant physical and ecological processes and the scale linkages between them. In coastal area, Hydrodynamic processes include: tidal currents, waves, and turbulence (Svendsen, 2002). Morphological processes include: sediment transport (bed load and suspended load, sand and mud), sediment sorting, bed level and bed form change, and mud effects. Ecological processes (represented in vegetation population dynamic module) consist of: species growth, spatial spreading, seed dispersion, competition, and mortality etc. Then the interactions between some critical processes are depicted. The model structure of generic geomorphological model shown in Section 3.6 is extended.

To verify the ecological processes and implementation, the model is applied to simulate the abundance of two types of freshwater macrophytes in Lake Veluwe, where the hydrodynamics is very quiescent in the lake. This makes the Lake Veluwe case a perfect case to verify the ecological processes in the model as the first step of the many more verifications. The model results have been verified against 4 years measured distribution of these two types of freshwater macrophytes in Lake Veluwe. And sensitivity tests are carried out to evaluate the relative importance of each ecological processes and the uncertainties from input parameters at the end of this chapter.

This chapter drafts the outline of a process-based modeling tool for multi-discipline (bio-ecology, morphology, water quality, hydrodynamics) research based on the generic geomorphological model.

5.2   Relevant processes and scales

Many researchers have raised the fundamental question of scale linkages over various earth sciences disciplines in the last decade (Baptist, 2005; Murray, 2007; Murray et al., 2008; D'Alpaos et al., 2007; Borsje et al., 2008; Bouma et al., 2007; De Vries & Borsje, 2007). As discussed in Chapter 2, De Vriend (2001) classified morphological processes as a hierarchy of linked scales. Similarly, Phillips (1995) and Klijn (1997) categorized ecosystems and the scale linkage of landscape problems using a hierarchical approach; Baptist (2005) reviewed scale linkages between the ecological processes and morphological processes in river floodplains. However, so far there has been little written about scale linkages within the field of coastal bio-geomorphology.

Relevant geomorphological processes and scales in coastal zone have been characterized in Section 2.2.1. Hereby the ecological processes and scales are described first.

5.2.1   Ecological processes and scales

Since the 1920's ecological models have been formulated that include biogeochemical cycles, photosynthesis, and growth of populations or individuals. Jørgensen et al.(2001) divided the ecological modeling approaches into 4 categories, i.e., conceptual model, static model, population dynamic model, dynamic biogeochemical models. This study focuses on the population dynamic model, where the state variables are numbers, size and structure, or biomass of individuals or species.

Lotka and Volterra developed the first population model, including growth, competition and decay processes, also known as the predator-prey model. Modern population dynamics models are basically extensions of the classic predator-prey model.

A population is defined as a collective group of organisms of the same species. Each population has several characteristic properties, such as population density (population size relative to available space), nativity (birth rate, spatial spreading, and seed dispersion), mortality (death rate), age distribution and growth forms. In particular, the population density is regarded as the status variable, which could be expressed as stems per square meters, or biomass per square meters (g/m^2). In the model, the properties are modified by a series of processes, i.e., growth, spatial spreading, competition, and mortality. However, not all processes are simple to represent in equations and some rules have to be integrated in processes as discrete constraints. Although the overall system of equations can hardly explain the full dynamics of population in nature (Jørgensen et al., 2001), the investigation of the equations still helps to improve our understanding of nature. The similar modeling approach is applied by Temmerman et al. (2005, 2007) in a loosely coupled modeling structure, while in this model, a built-in model is setup. Compared to Temmerman et al. (2007)'s model, two main advantages are identified:

1. This model has a formalized working stream, and the model structure is more standardized, flexible and extendible. Reusability of the knowledge is also much easier than Temmerman's approach;

2. This model has more capability. Two dimensional large scale effects are included in the model without losing accuracy, which widens the usage range quite significantly. Wave-vegetation interaction (Suzuki et al., 2011) and flexible vegetation effects (Dijkstra & Uittenbogaard, 2010) are ready to be integrated in this model soon.

Governing equation for population dynamic processes

As the analog of the advection-diffusion equation, the set of nonlinear governing equation for population dynamic could be:

$$\frac{dP_i}{dt} = rP_i\left(1 - \frac{P_i}{K}\right) + D_{veg,i}\left(\frac{\partial^2 P_i}{\partial x^2} + \frac{\partial^2 P_i}{\partial y^2}\right) \tag{5.1}$$

P_i is the state varible to be solved, i.e., the vegetation density for species i in unit of $[stem/m^2]$ for population density or $[gram/m^2]$ for biomass. The individual processes are discussed as the following.

95

Growth / mortality process

In general, there are two options for the population growth, i.e., the exponential population growth with the assumption of no resource limitation:

$$\frac{dP}{dt} = rP \tag{5.2}$$

or the logistic growth function in cases with resource limitation, which is adopted in this
study:

$$\frac{dP}{dt} = rP\left(1 - \frac{P}{K}\right) \tag{5.3}$$

Here t is the wall clock time, while the discretized time is denoted as $n\Delta t$. The analytical
solution in continuous form reads:

$$P_t = \frac{KP_0 e^{rt}}{K + P_0(e^{rt} - 1)} \tag{5.4}$$

or in discrete form:

$$P_{((n+1)\Delta t)} = \frac{KP_{(n\Delta t)}e^{r\Delta t}}{K + P_{(n\Delta t)}(e^{r\Delta t} - 1)} \tag{5.5}$$

where $\lim_{t\to\infty} P_t = K$ and $\lim_{n\to\infty} P_{(n\Delta t)} = K$. K is defined as the carrying capacity for the
species within the computation cell; r is the growth rate of the species; P is the population
density. The relative growth rate, $\frac{1}{P}\frac{dP}{dt}$, declines linearly with increasing population size
P. And the population at the inflection point (where growth rate is maximum), P is
exactly half the carrying capacity K.

In this equation, K and r are the parameters. Normally for one specific species to be
modeled, K is known before hand and is defined by input. So the growth rate r is the
only parameter for calibration in this model. r is a lump function of type of species,
temperature, salinity, bed shear stress, stress of competition, opportunity to get light,
nutrients and so forth.

Furthermore, the stability of this nonlinear system depends significantly on the growth
rate r, which had been discussed in many literatures like May (1976), etc. When $r < 0$,
the population will eventually die, independent of the initial population, which implies

decay of the system. When $0 < r < 1$, the population will be stable at $(r-2)/(r-1)$ gradually, also independent of the initial population, if the environment conditions are favorable, and simulation time is long enough, other constraints are not applied. When $r > 1$, the system will oscillate between 2 values, 4 values with higher r, and eventually lead to chaos, later on diverge for almost all initial values (Gershenfeld, 1999; Tsoularis, 2001).

If we consider the whole range of biological activity, we again observe a wide range of time scales. It is estimated that (Jørgensen et al., 2001):

- for bacteria, such as Escherichia coli, the rate of increase is around 60 per capita/per day and mean generation time is the scale of 0.014 days, while

- for algae, the mean generation time is scale of hours, and

- for insects, the rate of increase is around 0.01 per capita / per day and mean generation time is the scale of 100 days, and

- for mammals, the rate of increase is around 0.01 per capita / per day and mean generation time varies the scale of 100 days to 1000 days.

The mean vegetation generation time is typically seasonal or annual. Furthermore, the development of spatial pattern of vegetation is generally in the scale of years to decades (Li, 2009), which is notably at the same range of the middle scale of morphology change. At certain scales, the physical environment has significant effects on the growth and decay processes which have been included in models by additional rules.

Spatial spreading

Spatial spreading, similar to diffusion process (Eq.5.6), is crucial for the development of the spatial pattern. The spatial pattern, extracted from GIS map, satellite image and aerial photograph, can be used to validate a vegetation population spatial spreading. Spatial extension processes may happen within the growth scale, or much faster than the growth scale (Sanz, 1999, 2000), which add the complexity to the modeling practice. We might parameterize the stochastic dynamics in the model, which is not trivial as well.

$$\frac{dP}{dt} = D_{veg}\left(\frac{\partial^2 P}{\partial x^2} + \frac{\partial^2 P}{\partial y^2}\right) \tag{5.6}$$

Competition/interaction among species

Two (or more) plants may influence each other by means of competition and coexistence. Different species of vegetation compete for nutrients, space, light etc. For vegetations the competition is mostly often governed by local processes, but for animals the competition takes place over larger domains. In a model this process often implemented as a formulation limited by the interaction of a relatively small number of computational segments in space and in a few steps in time. For example,

$$\frac{dP_1}{dt} = p_c[(P_1 - K_{P_1}) + (P_2 - K_{P_2})] \tag{5.7}$$

Where: P_1, P_2 are the population density of the first and second species;
K_{P_1}, K_{P_2} are the carrying capacities for both species;
p_c is the probabilistic coefficient, which needs calibration.

Seed dispersion

The dispersion of seed determines ecological characteristics of plants and their communities (Thompson et al., 2008). Even though different species have their own dispersal strategies, seed dispersal is broadly characterized by a canonical length scale over which seeds move and by a dispersal vector that transports them. The length scales of dispersal vary from highly localized ($< 1m$) to long distance ($1 - 100m$) (Thompson et al., 2008; Trevethan et al., 2008). Ecologists (Van Koppel and Temmerman, personal communication, 2011) suggest that it is quite realisitic to assign equally a random distribution of seeds for all over the study domain in coastal zone.

$$\frac{dP_1}{dt} = f(P_1, \quad P_2, \quad average\ age\ in\ this\ cell, \quad age\ in\ the\ neighbour\ cells) \tag{5.8}$$

5.2.2 Coupling of scales

Many researchers have argued that the influence of contingent or emergent behavior determines the scale of the problem (Murray, 2003; Werner, 2003, etc.). So the small scale problems are those whose time and space scales are small enough that temporal variations can be effectively averaged and fixed constitutive equations, which can be deduced by flume experiments and field measurement. The large scale problems are those whose dependent or evolving condition can be specified explicitly by initial/boundary condition or governing equations (Wilcock, 2003). In each discipline, a classification of hierarchical scales could be carried out (concepts of Odum and De Vriend). For both 2 types of modeling approaches (refer to Section 1.1): the "explicit numerical reductionism" ("bottom-up" approach), and the "top-down" (Murray, 2003; Murray et al., 2008;

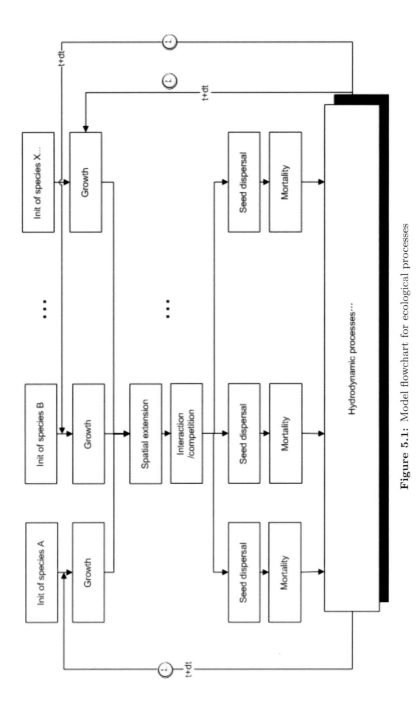

Figure 5.1: Model flowchart for ecological processes

Werner, 2003), respectively, the scale coupling is implemented by parameterization or by incorporation of a few crucial aspects of dynamics from processes at the next level of (larger or smaller) scale. This argument provides the rationale for the integration of ecological processes into the generic geomorphological model.

Furthermore, De Vries and Borsje (2007) argued that there exists a shortcut for scale coupling: large scale climate change can directly influence short-term morphodynamic via its impact on zoo benthos populations (refer to Figure 5.2), which still needs more measurements to support. In this study, both ecological, main vegetation, and morphological processes at the middle scale are focused on. In the following section, the interactions between the processes are discussed.

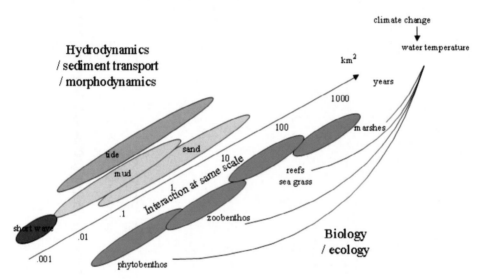

Figure 5.2: Coupling of morphological and ecological processes and possible shortcuts (Modified from De Vries et al., 2007)

5.3   Interactions between processes

In this section, the interactions between morphodynamic processes and ecological processes have been overviewed. On one hand, the presence of vegetation might slow down the current speed, thus may cause sedimentation (Baptist, 2005). On the other hand, flow around vegetation stems and leaves produce extra local turbulence, thus may cause scour hole around the stems and enhance sediment suspension by "monami", a hydrodynamically induced synchronous waving of sea grasses leaves (Ghisalberti & Nepf, 2002). This might influence the larval mussel as well (Grizzle et al., 1996).

Existence of vegetation damps wave height and attenuates wind waves. Dense vegetation stems and the root systems might also introduce strong resistance against erosive force to some depth under the water-sediment interface, but might increase the possibilities of erosion in soil patches under extremely high near-bed shear stress during storms. Strong hydrodynamic environment would also affect the growth and spread of vegetation (Schanz & Asmus, 2003).

5.3.1 Vegetation Impacts on flow

At large spatial scale, the presence of vegetation introduces additional roughness by drag force, and also introduces extra turbulence around the stems and leaves, which might be considered at small scale. These phenomena have been studied by many researchers.

At large scale, usually the drag force of vegetation is represented by change of velocity profile and extra bed roughness. Different approaches to describe the velocity profile include:

- Based on the traditional logarithmic velocity profile for turbulent boundary layers, an offset from the bottom is included.

- Mixing layer model. Ghisalberti and Nepf (2002) purposed a hyperbolic tangent profile based on laboratory studies. Baptist (2005) suggested a double exponential velocity profile.

In this study, Baptist (2005) method is applied for the 2d option, and the velocity profile is elaborated as bed roughness described as:

- For nonsubmerged vegetation ($h < k$):

$$C = \frac{1}{\sqrt{\frac{1}{C_b^2} + \frac{C_D n h}{2g}}} \tag{5.9}$$

- For submerged vegetation:

$$C = \frac{1}{\sqrt{\frac{1}{C_b^2} + \frac{C_D n h_v}{2g}}} + \frac{\sqrt{g}}{\kappa} \ln(\frac{h}{h_v}) \tag{5.10}$$

Where:
k is defined as the effective height of vegetation (m);
C_b is defined as Chezy bed roughness without vegetation $(m^{0.5}/s)$;
C_D is defined as the drag coefficient of vegetation (-);
m is defined as cylinder density per unit area (m^2);
D is the cylinder diameter (m);
h is the water depth (m).

At small scale, the 1D vertical model with the extend $k - \varepsilon$ model of Uittenbogaard (2003) is applied in this model, i.e., an extra sink term of drag force is added in the momentum equations and extra source terms are introduced into the transport equation of k and ε equations. The formula is listed as Equation 4.1, 4.5 and 4.8 in Uittenbogaard (2003), and cited as Equation 1, 2 and 5 in Temmerman et al. (2005).

The 3D approach would provide more detailed information in the hydrodynamics and coherent structures in turbulence around the vegetation stems. But it would be more computational expensive, which limits its application range. For instance, for mid-term morphology study in salt marsh, the heavy computation efforts make it almost impossible through the 3D approach. The 2D approach generally is less complex in computation, while it requests more knowledge to have reasonable parameterization. For hydrodynamics, Baptist (2005) used a data-driven approach, i.e., using the 3D approach to generate a huge hydrodynamic and vegetation in datasets, then using the datasets to evaluate the parameters in his 2D approach. Thus, the 2D approach is equivalent to the 3D approach for hydrodynamic computation, but this doesn't hold for sediment transport and morphology, which is discussed below.

5.3.2 Vegetation Impacts on waves

With existing of vegetation in the coastal salt marsh, wave energy is dissipated due to the enhanced bed friction and drag forces. Dalrymple et al. (1984) made estimation on the wave dissipation over rigid cylinders by integrating the force on a cylinder over its vertical extent. This was extended by including varying depths and the effects of wave damping due to vegetation and wave breaking for narrow banded random waves (Mendez & Losada, 2004). Suzuki et al. (2011) describes the wave dissipation over a vegetation field by the implementation of the Mendez and Losada formulation in a full spectral model SWAN, with an extension to include a vertical layer schematization for the vegetation. An extra sink term is introduced in the wave energy equation, thus the wave energy dissipation caused by vegetation is taken into account in solving the wave action equation. Refer to Equation 6.9 of Suzuki (2011).

5.3.3   Vegetation Impacts on sediment transport

Vegetation impacts on sediment transport in two ways. For suspended sediment, the vertical diffusivity profile and bottom boundary thickness are modified. The settling of the sediment particles is expected also to slow down by the vegetation.

For 2D simulation, the vertical diffusivity profile is assumed as constant-parabolic-constant, similar to the form given by Van Rijn (1993) without vegetation. With presence of vegetation, the diffusivity profile is modified as a linear interpolation within the vegetation zone. For the symbol definition refer to Equation 4.7 in Section 4.2.3.

$$D_V = \alpha_{veg} * \kappa \beta u_* h \qquad\qquad z < z_a \tag{5.11}$$
$$D_V = \kappa \beta u_* z(1 - z/h) \qquad z_a < z < 0.5h \tag{5.12}$$
$$D_V = 0.25\kappa\beta u_* h \qquad\qquad z \geq 0.5h \tag{5.13}$$

Where: α_{veg} is a calibration coefficient, limited to $10 < \alpha_{veg} < 100$. β is defined as:

$$\beta = 1 + 2\left(\frac{\omega_s}{u_{*,c}}\right)^2 \tag{5.14}$$

and limited to the range $1.0 \leq \beta \leq 1.5$.

For 3D simulation, the profile is computed from $k - \varepsilon$ model described as Uittenbogaard (2003), with small modification of the const coefficient C_μ (Lopez & Garcia, 1998).

For bed load, sediment erobility is expected to be hampered by vegetation stems. It is assumed the critical shear stress increases as a function of the vegetation density. And the availability of sediment erosion is a function of the occupation of the root system. The book keeping system in the bed state stores the root info varying with time.

5.3.4   Ambient environment impacts on Vegetation dynamics

The interaction of flow with vegetation dynamics has been shown to influence productivity (Schanz and Asmus 2003), growth of root system (Peralta et al. 2006), uptake of carbon and nutrients (Cornelisen and Thomas 2006; Morris et al. 2008), seed dispersal pattern (Thompson et al., 2008), etc..

In this study, these sorts of impacts are highly simplified into a set of parameters and rules. It is assumed that:

- Bed shear stress
  If ambient bed shear stress around the vegetation stems is higher than a critical value, the growth rate of the vegetation decreases.

- Inundation depth and time
  If inundation depth is too high or the inundation time is too long, the growth rate decreases.

- Sediment concentration
  If sediment concentration in the water column is too high, the secchi depth deceases and thus the growth rate decreases.

- Temperature and temperature gradient
  If temperature is low, the growth rate decreases. If the temperature gradient is negative, the growth rate might decrease gradually to negative as well. This rule could ensure the seasonal variance of vegetation. In spring, the growth rate is positive, while in autumn, the growth rate gets decrease. This rule is also helpful for highly spatial variance of vegetation species in estuaries and other coastal zones.

- Salinity
  Some vegetation species could bear higher salinity than others. This rule makes sure that different species survive in their favorable environment.

- Wave force
  The drag force induced by wave is the predominate force on submerged vegetation stems and leaves in shallow water. Strong wave force might have negative effects on the vegetation growth. Vegetation caused wave energy dissipation is taken in account in the wave model. Suzuki (2011) is applied in this study.

The definition of the parameters listed above are very preliminary, and need elaborate calibrations for specific species and study sites. Furthermore, the effects of flexible vegetation are not taken into account in this study. Refer to Dijkstra and Uittenbogaard (2010) for more information on the effects of flexible vegetation.

5.3.5   Discussion on the morphological factor

The morphological factor is to accelerate the morphological processes. It has been extensively discussed for morphodynamic processes by Roelvink (2006); Lesser (2009). Discussions on mass conservation are also carried out in Section 3.4.4. When ecological processes are involved, concerns are arising from the application of morphological factor.

Tide dominated case
Morphological change are determined not only by one process rather by a complex balance between several processes (Lesser, 2009). To bring the gap of scales of hydrodynamic

processes and morphological processes, some acceleration techniques are applied. To choose a representative morphological tide is the first step. Based on literature review and numerical experiments, Lesser (2009) suggested 1.08 times the main M_2 tide constituent plus a combination of diural signal $C_1(=\sqrt{2O_1 K_1})$ as the morphological tide ($= 1.08 *$ $(M_2 + C_1)$). And the period could be simply double the M_2 tide (1490.47 minutes). And the simulation should be simulated for at least the duration of a spring-neap cycle. In the tide dominate case, the morphological factor could be 10 to 400 (Roelvink, 2006; Van De Wegen, 2010; Van De Wegen et al., 2008).

Wave involved case
In coastal zone, wave is an important process. For long term morphological simulation, the wave input reduction could be determined by estimating the morphological impact of different wave class(Lesser, 2009).

The CERC formula to estimate the "morphological impact" M_c and "morphological scatter" in wave class are:

$$M_c = p_c H_{s,rep}^{2.5} = \frac{\sum_{i=1}^{N_c} H_{s,i}^{2.5}}{N_s} \tag{5.15}$$

$$\Delta_m = M_c \frac{\sum_{i=1}^{N_c} \sqrt{[|H_{s,i} - H_{s,rep}|^{2.5}]^2 + [\Theta_i - \Theta_{rep}]^2}}{N_c} \tag{5.16}$$

Where:
N_c: number of wave records in the wave class;
N_s: number of wave records in the season;
$H_{s,i}$, Θ_i: Individual significant wave height and direction;
$H_{s,rep}$, Θ_{rep}: Representative significant wave height and direction in the wave class.

The wave classes could be estimated by numbers of wave records in a scattering points chart of wave height (H_s) against wave direction (Θ) (Figure 5.7 of Lesser (2009)). We can also see that the two factors listed above have high correlation of the probability of occurrence falling in the wave class.

Once the wave classes N are determined, the average morphological factor is: $(365 \times 24 \times 60) \div (N \times 745)$. For each wave class, the morphological factor could be determined as (Lesser, 2009):

$$f_{morfac} = \frac{p_c \times seasonal\ duration}{T_{morphotide}} \tag{5.17}$$

$$= \frac{occurence\ days\ in\ the\ season \times 24 \times 60}{745.24min}$$

In serial online simulation, to avoid the shock transition when wave condition changes, before each new wave condition is applied, Lesser (2009) also suggested to have a hour period to shut down the morphological change and to leave the hydrodynamics adapting to the new wave forcing.

In parallel online simulation, an averaged morphological factor could be used. The ecological affects are taken into account individually in each wave condition. The "mormerge" function would sum up the morphological change based on the probability occurrence of each wave condition.

Ecological processes involved case

If ecological processes are involved, special care should be taken for the morphological factors. Typical coupling time step of ecological processes and morphodynamical processes are days, weeks, and months.

For tidal dominate case, the morphological factor may stay the same. For example, if we set hydrodynamic time step 0.1 minute, morphological factor of 30 for a one-year morphological simulation, and ecology-morphology coupling every week.

Table 5.1: Acceleration factor

	Hydro. modeling	Mor. modeling	Eco. modeling
Time step	0.1 [min]	$0.1 \times 30 = 3$ [min]	$\frac{(7 \times 24 \times 60)}{30 * 0.1} = 3360$ [min]
Total simulation steps	$\frac{12 \times 60 \times 24}{0.1} = 172800$	$\frac{12 \times 60 \times 24}{0.1} = 172800$	$\frac{12 \times 60 \times 24/0.1}{(7 \times 24 \times 60)/30 * 0.1} \simeq 52$
Factor	1	30	$\frac{(7 \times 24 \times 60)}{30 * 0.1} = 3360$

Thus for the morphological year, 24 tidal cycles are forced. The hydrodynamic simulation period is 12 days. For ecological simulations, $(12 \times 60 \times 24/0.1) \div$ (ecological time step) $\simeq$ 52 times, i.e., once a week. The ecological time step is set as the communication interval between ecological modules and morphodynamic modules.

The growth rate r is defined with unit [1/time], e.g., [1/week]. The values are usually estimated based on experiments and expert's judgment. However, a similar acceleration factor could be included in the input growth rate to speed up the whole simulation loop. Furthermore, it is noticed that in tidal environment, higher coupling frequency is usually essential. Otherwise it is risky that the input ambient physical factors for the ecological model and the ecological effects on physical processes would be under-sampled. A computational example has been shown in Section 5.5.4.

For wave involved case, in case of parallel online simulation, when "mormerge" function is used, the similar approach is used to determine the ecological step, while in serial online simulation as Lesser (2009) purposed, time varying morphological factors are used in the simulation, thus the ecological steps are varying during the season as well by the same formula in Table 5.1, i.e., $(7 \times 24 \times 60)/($hydrodynamic time step$\times$ morphological factor). Thus time varying communication intervals between ecological modules and

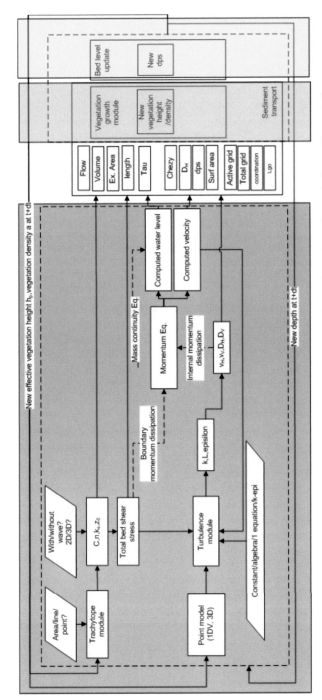

Figure 5.3: Data flow diagram for the bio-geomorphological model: Bed roughness (refer to Figure 3.8)

morphodynamic modules are possibly applied, even though this function is not yet in this system.

5.4   Extended model Structures and Dataflow

The ecological processes are also integrated into the bio-geomorphological model through Open Processes Library in the frame of Delft3D-WAQ.

Although this is not strictly necessary, the model setup was chosen such that the Delft3D-FLOW provides flow information at each time step, e.g. water level, depth, velocity vector etc. to the WAQ modules. Sediment transport is modeled separately as bed load and suspended load using one of the implemented formulations (Van Rijn 1993). The suspended sediment is modeled using an advection-diffusion equation with the source and sink term as given by Lesser et. al. (2004). The sediment sorting in vertical direction is tracked by a book-keeping system. After the bed load transport vectors and the suspended load source and sink terms have been computed, the bed level is updated and accelerated (Roelvink, 2006). Subsequently, the new bed level is transferred back to flow module and vegetation growth module. At the same time, with the flow information, the vegetation growth module starts. The ecological processes implemented in the system include vegetation growth, spatial spreading, species competition, mortality, seed dispersal. The vegetation growth module provides vegetation height and spatial density to the vegetation effect module. The vegetation effect module computes the impact of the vegetation on the (either 2D or 3D) flow, e.g. enhanced bed roughness, bed shear stress or increased turbulence. These quantities are then passed back to flow solver at a user-defined time interval, or at the time steps of vegetation growth model. The vegetation effects are determined based on a modified version of improved bed roughness approach of Baptist (2005) for large scale depth-averaged model or the 3D k-epsilon approach of Uittenbogaard (2003) for small scale properties, e.g. vertical variations, 3D flow and turbulence due to vegetation, etc.

Through this simultaneous (online) coupling approach, the effect of vegetation to flow structure and the flow effects to vegetation growth (population dynamics) are modeled. The data flow of extended bed roughness is demonstrated in Figure 5.3.

5.5 Validation of ecological processes

5.5.1 Introduction

The vegetation population dynamics processes are validated by reproducing the spatial pattern of two species in a semi-closed inland lake, Lake Veluwe, a artificial lake located northeast of Amsterdam. The lake has a surface area of 30 square kilometers and average water depth of 1.5 meter (Figure 5.4). In late 1960's, severe eutrophication leads to a big shift of vegetation from macrophytes to only sparse patch of submerged *Potamogeton Pectinatus* (abbreviated as *Pp*). To remediate the water quality problem, measures has been taken by reducing phosphorus load and introduce Charophyte meadows (*Chara aspera*, abbreviated as *Cs*) from 1980's. Nowadays, the persistent Charophyte meadows are observed and the underwater light climate has improved. More background information and the restoration project could be found in Van Den Burg (1999).

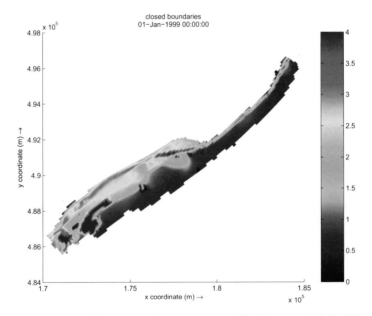

Figure 5.4: Model domain, grid arrangement and bathymetry in Lake Veluwe

Pp and *Cs* are regarded as water quality indicators for fresh water body. They co-exist in water with good quality and their growth depends on light, nutrients and optimal inundation depth, etc. (Li, 2009). The biological properties of *Pp* and *Cs* are also described in Chapter 7 of Li (2009), from which the brief introduction is cited in the following Table 5.2:

Table 5.2: General information about the species
(Pp. (*Potamogeton Pectinatus*) and Cs (*Chara aspera*)) (from Li, 2009)

Parameter	Values	
	Pp	Cs
Max density	600 $[stem/m^2]$	600 $[stem/m^2]$
Max height	2 $[m]$	0.22 $[m]$
Max Diameter	0.01 $[m]$	0.01 $[m]$
Growth rate	0.5 $[1/week]$ (at 15°)	0.7 $[1/week]$ (at 15°)
Life span	multi year	about 3 months
Seeding strategy	mainly by roots	oospores and roots
Root	strong, big	weak, small
Predation	ducks, fishes, birds	ducks, fishes, birds
Inundation depth	4 $[m]$	2 $[m]$
Salinty	3 $[ppt]$	20 $[ppt]$
Shear stress	2 $[Pa]$	0.26 $[Pa]$
Light	no limitation	need high light availability
pH	worse grow with high pH	better grow with high pH
Water quality	indicator for good water quality	indicator for good water quality

5.5.2 Model setup

The model covers the Lake Veluwe by around 60 × 60 meter grid cells. The maximum depth is around 4 meters, while bathymetry of almost half of the lake area is less than 1 meter (Figure 5.4). Since the lake is a closed water body, the driven hydrodynamic forces are wind and very little controlled inflow and outflow discharge points through locks and weirs around the lake. Thus the hydrodynamics and morphodynamics in this closed environment are mild, which makes it a proper case to validate the ecological processes in the model.

5.5.3 Results

Measurement of the amount of populations for macrophytes, and spatial pattern include GIS maps from 1994 to 1999. The simulated spatial patterns from year 1994 to year 1997 are shown in Figure 5.5.

To access the simulated spatial patterns of vegetation species, an analog of Brier Skill Score (BSS), PoD and TS index are applied (the definitions refer to Section 2.2.3 and Li (2009)). The modeled abundance of the species Cs is in occupation percentage, which is classified into 7 classes (Table 5.5) and then compared to the field measurements.

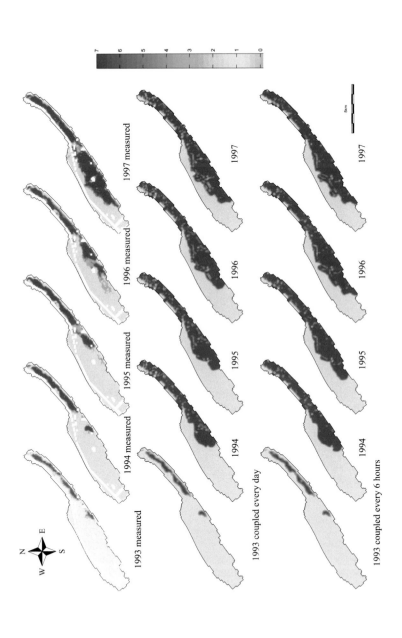

Figure 5.5: Upper Panel: Measured *Chara aspera* population densities and distributions from year 1993 to year 1997. Middle Panel: Simulated *Chara aspera* spatial pattern from year 1993 to year 1997(from left to right). Lower Panel: Simulated *Chara aspera* spatial pattern from year 1993 to year 1997, when coupled every 6 hours, to check timesteping of r. The population density is classified from low to high into Class 0 to Class 7.

111

Table 5.3: Parameter lists and ecological model settings
(Pp. (*Potamogeton Pectinatus*) and Cs (*Chara aspera*))

Parameter	Values	
	Pp	Cs
Max density	600 $[stem/m^2]$	600 $[stem/m^2]$
Max height	2 $[m]$	0.22 $[m]$
Max Diameter	0.01 $[m]$	0.01 $[m]$
Growth rate	0.5 $[1/week]$ (at 15°)	0.7 $[1/week]$ (at 15°)
Life span	10 [years]	about 3 [months]
Lateral expansion	5 stems per week if 2 over 8 neighbouring cells are higher than 50%	25 stems per week if 2 over 8 neighbouring cells are higher than 50%
Seeding strategy	-	a random number (0-5%)
Inundation depth	4 $[m]$	2 $[m]$
Shear stress	2 $[Pa]$	0.26 $[Pa]$

Table 5.4: Parameter lists and morphodynamic model settings

Module	Parameter	Value	Description
Flow	Δt	5 [mins]	Computational time step
	T	1 [year]	Hydrodynamic simulation time period
	v_H	0.1 $[m^2/s]$	Horizontal eddy viscosity
	D_H	10 $[m^2/s]$	Horizontal eddy diffusivity
Wind	Δt	60 [mins]	Hourly data measured at Lelystad station
	$\Delta t_{ecostep}$	1 [day]	Ecological simulation time step

PoD and Ts are the indicators for spatial patterns, defined as:

$$PoD = \frac{Ac}{Ac+Am} = \frac{Hits}{Obs} \tag{5.18}$$

$$TS = \frac{Ac}{Af+Ao-Ac} = \frac{Hits}{Forecast + Obs - Hits} \tag{5.19}$$

Where: the $\langle , \rangle$ denote arithmetic means. A_c is the correctly predicted area; A_m is the miss shooting area; A_f is forecasting area and A_o is the observation area. The criterion of correctly predicted is set as 15 percent, e.g., if both the measured and the modeled species occupation in one cell are more than 15% (≥ 3 in class), we say then the pattern in this cell is correctly predicted by the model.

The index PoD equals 1 means that the observed vegetated area is inside the model predicted area or equal. PoD index shows how good the predicted vegetated area is covered by the observed area. However, the model might overestimate the vegetated

Table 5.5: Classes of the Occupation of the vegetations

Class	Occupation percentage
0	0 %
1	< 1 %
2	1 - 5 %
3	5 - 15 %
4	15 - 25 %
5	25 - 50 %
6	50 - 75 %
7	75 - 100 %

area where there is not vegetation by observation. The index TS equals 1 means that the observed vegetated area and the predicted vegetated area are perfectly overlayed, even though also the predicted vegetated area could be larger than the observed area.

The Brier Skill Score is focused more on each cell rather than spatial pattern. The simulated vegetation percentage is compared with observation cell by cell. The Brier Skill Score is defined as:

$$BSS = 1 - \frac{\langle (\Delta P_{modeled} - \Delta P_{measured})^2 \rangle}{\langle (\Delta P_{measured})^2 \rangle} \tag{5.20}$$

If BSS is or close to 1, the simulated vegetation occupation match field measurement in each cell perfectly, where the spatial pattern doesn't have to be right. In this model, PoD index in Table 5.6 are higher than 94%, which indicates that the model results complete cover the observed vegetation distribution pattern. The BSS index shows the same trend in 1994 and 1997, even though in 1995 and 1996, the index values are a bit low. BSS index gives out quantitative analysis by comparing the model results and field observation in each cell. Mismatch of vegetation abundance and phase in several segments (grid cells) might lead to low BSS index value, such as the index value in 1995 and 1996, which also demonstrates the difficulties to represent the complex interactions of ecological processes quantitatively. The mismatch might be due to some ecological parameters not set correctly or even some processes being missed, e.g., the seed bank development in the bed is crucial for the Cs species to extend, thus from the measurements, in the first two years, from 1993 to 1994, expansion of the vegetation patches are hardly observed. In this aspect, we need more collaboration with ecologists to improve the ecological processes in the model.

In general, both quantitative analysis (Table 5.6) and visual examination (Figure 5.5) illustrate that the spatial distribution pattern from the model reasonably represents the field observed pattern of the vegetation species at Lake Veluwe, though the growth rate and/or expansion rate seem too high in the middle period.

Table 5.6: Spatial pattern indexes for the modeled vegetation (Cs) distribution

Year	1994	1995	1996	1997
PoD	0.94	0.94	0.96	0.96
Ts	0.32	0.38	0.44	0.74
BSS	-1.2	-0.71	-0.36	0.62

5.5.4 Discussions

Coupling interval

First of all, the formulations of population dynamics model w.r.t. timestepping is checked. The growth rate r should be independent of time step/coupling interval and has dimension 1/time. Hereby we run the model again with the same settings except the coupling time. The vegetation population results from a model coupled every day should be similar to the results from a model coupled every 6 hours. The lower panel in Figure 5.5 showed almost identical pattern to the middle panel in Figure 5.5. To quantify the difference, the results from the model coupled every 6 hours are compared against the measured data using the PoD, TS and BSS in Table 5.7.

Table 5.7: Spatial pattern indexes for the modeled vegetation (Cs) distribution, coupled every 6 hours

Year	1994	1995	1996	1997
PoD	1.0	0.9922	0.93	0.71
Ts	0.42	0.50	0.55	0.57
BSS	-0.48	-0.12	0.04	0.44

The indexes show small differences between the model coupled every day and the model coupled every 6 hours. The difference might origins from stochastic properties included in the ecological processes. Random numbers are used as parameters in the model to represent the stochastic properties. However, results from both models showed the same trend with time: PoD stays close to 1; Ts and BSS get better with time (Table 5.7).

Sensitivity of parameters of the ecological model

To examine the sensitivity of the model responding to the model parameter settings, sensitivity tests have been carried out for the crucial parameters, such as growth rate r, critical inundation depth, critical shear stress. Values of the parameters are given a variation from ±5%, ±10%,±20% to +50%.

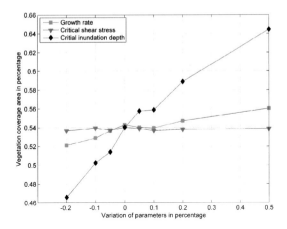

Figure 5.6: Sensitivity test: parameters

The sensitivity test on parameters shows that the inundation depth is the most sensitive parameter (Figure 5.6). 20 % of change of inundation depth leads to 20 % of difference of vegetation coverage area. It seems there is no significant difference of vegetation coverage area by increasing of the growth rate, while the vegetation coverage area is more sensitive by decreasing the growth rate. The model results also show that the critical shear stress is not very sensitive parameter. These conclusions are in line with the previous study, for instance, Li (2009).

Sensitivity of ecological processes

To examine the sensitivity of the model responding to the ecological processes, sensitivity tests have been carried out by turning off certain processes, such as, growth, lateral diffusion, competition, seed dispersal and mortality. The results show that mortality and interaction between species are the most important processes (Figure 5.7).

5.5.5   Conclusions

Validation on the implementation of ecological processes and their interaction are focused on in this section. An ecological model for Lake Veluwe has been set up with a stagnant environment with very little morphodynamic forces. Even though, very little effort has been put into the model calibration, the observed spatial patterns of two macrophyte species in the lake have been well captured by the model. The sensitivity analysis on

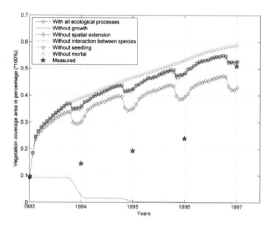

Figure 5.7: Sensitivity test: the ecological processes

the parameters showed that the ambient physical environment is crucial for the lateral expansion of certain macrophytes. The sensitivity analysis on the processes shows that the most sensitive processes are seed dispersion, which is in line with the conclusions from literatures (Li, 2009).

Even though most of the parameters and definition of the ecological processes are rather instinctive, and the interactions between species are far away from being accurately quantified, the model still gives out reasonably results, which shows the model has great potential to be a process-based bio-geomorphological research tool.

5.6   Discussions and conclusions

In this chapter the possibilities have been explored to integrate the ecological processes and the interactions between them into the generic geomorphological model frame described in Chapter 3. This model extends the modeling approach of Temmerman et al. (2007). To extend the generic geomorphological model to be a bio-geomorphological model by integrating ecological processes and the interactions with geomorphodynamic processes, the scales of ecological processes are identified. A classification of hierarchical scales is made for ecological processes. The interaction of ecological processes and morphodynamic processes are listed by integrating up-to-date existing knowledge. The model structure and data flow described in Section 3.6 are modified to adapt the integration processes.

The validation case for Lake Veluwe shows the capability of this bio-geomorphological modeling tool. The ecological processes are reasonably reproduced and the evolution of spatial pattern of macrophytes in the lake is well captured by the model, compared to the measured data from 1993 to 1997, and also to the results from other models, such as, the cellular automata model and multi-agent-based model (Li, 2009).

In the next chapter, capabilities of the integrated model are further explored by means of three applications at different temporal and spatial scales: a schematized tidal basin, a salt marsh restoration project in the US, and a dynamic coastal zone in the East China Sea.

Chapter 6

Capability of the

bio-geomorphological model

6.1   Introduction

After the development of the geomorphological model in Chapter 3, and model validation of the geomorphological processes in Chapter 4, Chapter 5 extended the generic geomorphological framework to a bio-geomorphological model by including ecological processes. In this chapter, the focus will be on three applications of this integrated model at different temporal and spatial scales: a schematized tidal basin, a salt marsh restoration project in the US, and a dynamic coastal zone in China.

Section 6.2 uses the new bio-geomorphological modules to simulate the effect of vegetation on the morphodynamics of a schematized tidal inlet similar to those studied by Roelvink (2006); Tung et al. (2009); Dissanayake et al. (2009). The analysis focuses on the effects of vegetation on the channel network development in the tidal basin and the overall characteristics of the tidal basin. Subsequently, Section 6.3 applies the new modules to a salt marsh restoration project in Puget Sound near Seattle in the northwest of the US. This study focuses on the sedimentation within the salt marsh and the river channels, and the development of the spatial patterns of vegetation after dike removal (including multi-species interaction). Finally, Section 6.4 focuses on the Jiangsu coastal zone in the southern Yellow Sea, China between the Yellow River and the Yangtze River.

This area is an important, dynamic ecological area (see e.g. Ke et al. (2009), and Zuo et al. (2009)) with unique morphological characteristics in the form of large, shoreface connected radial shape sand ridges. Although the intention was to test the full bio-geomorphological system on this study area, time did not allow for it. However, rather than completely excluding this study area from this thesis, we include this first step since our simulations (carried out using standard Delft3D online MOR) shed some new light on the genesis of the sand ridge pattern as it exists today.

6.2 Case 1: Bio-geomorphological modelling in salt marsh

6.2.1 Introduction

Tidal salt marshes are the marine intertidal areas with soft substrate embracing channels and shoals, which are predominantly colonized by various vegetation species. The high economical and ecological value of tidal salt marshes is getting increasing awareness in recent decades. The channels on the tidal salt marshes serve as the main paths for flow, sediment, nutrients, etc. (Temmerman et al, 2007). The formation of the channel networks is determined by the interactions of tide-induced pressure gradients and residual currents, wind-driven waves, swell, sand/mud transport, sediment sorting in horizontal and vertical direction. The presence of vegetation adds to the complexity. Existence of vegetation does not only slow down the flow and cause sedimentation, but also increases the local turbulence by flow around the stems and leaves, and causes erosion. The root systems introduce strong resistance against erosive force to some depth under the water-sediment interface, but might increase the possibilities of erosion in soil patches under extremely high near-bed shear stress during storms. Many efforts have been put into these topics and several interesting hypotheses have been made. For instance, Temmerman et al. (2007) claimed that the dynamic vegetation patches obstruct the flow, leading to concentrated flow and channel erosion. Their opinion is new compared to the traditional view that vegetation would reduce erosion of channels in fluvial and tidal environments. Nevertheless, their conclusions were drawn through a loosely coupled model where the vegetation dynamic effects on morphology evolution within one year are neglected. Within a year, morphology with the tidal marsh develops with constant vegetation information (patch size, stem height, etc.), which is only updated once a year. In this section we hypothesize that the dynamic vegetation effects cannot be ignored for the tidal salt marsh channel formation. To analyze the crucial effects of dynamic vegetation on channel formation on a tidal salt marsh at various temporal (spring-neap cycle, seasonal, annual, decadal) scales, the process-based, tightly coupled, generic geomorphological model developed in this study is employed to hindcast the channel formation.

The prototype of the case is a salt marsh to be restored in San Francisco Bay. The study area is a 400 m × 700 m, bare sandy mild-slope coastal tidal basin. It was used as a dredging sediment dumping site initially. Afterwards the sea side bank of the basin was opened and the area is supposed to be restored as a natural salt marsh. Figure 6.1a shows what the basin looks like now. This type of salt marshes can be found in many places. To generalize the problem, a schematized model is set up to identify the dominant processes and main morphological features in this section. The pioneer vegetation species Spartina spp. (cordgrass) is chosen as an example to demonstrate the ecological effects on the channel formation in the salt marsh. Developing patterns of the ebb tidal delta, flood delta in the salt marsh over a timespan of decades will be investigated. Effect of the vegetation population dynamics on the channel developing rate onshore at various time scale (spring-neap cycle, seasonal, annual, decadal) will be discussed.

6.2.2   Model setup and dominant processes

A schematized model is set up to demonstrate the dominant processes. The grid arrangement and the initial bathymetry are shown in Figure 6.1b. The model covers an area of 1.8 × 1.5 kilometers discretized into 30 × 30 meter grid cells. The basin covers area of 1.12 × 1.5 kilometers. The initial bathymetry inside the basin is uniform (-1.5m), and outside of the basin linearly increases to -5 meter. The only open boundary is forced by a harmonic tide from seaside in the North. The tide range is 2 meters. A sand bank is set as a sediment source in the model domain. The initial cross section area of the tidal inlet is 135 m^2 (i.e., 90 meters (3 grid cells) wide by 1.5 meters deep). Initial volume of ebb delta is 0, while the initial tidal prism volume is then 3.36Mm^3 (=1120*1500*2), and the initial intertidal area is then 0 as well. With morphological changes, the cross section area of tidal inlet, sediment imports and exports are expected to increase till equilibrium, if there exists equilibrium. Other morphodynamic parameter settings are listed in Table 6.1.

The vegetation species used here is a schematized species similar to Spartina. Spartina, commonly known as cordgrass or cord-grass, is widely distributed all over the world. They form large, often dense colonies, particularly on coastal salt marshes, and grow so quickly that some of them are regarded as invasive species. The species vary in size from 0.3 m to 2 m tall. The zonation of spartina can expand by root-rhizome system and seedling (refer to http://plants.usda.gov/ for details). The ecology related parameter settings are listed in Table 6.2.

With respect to morphological change for the tidal salt marsh channel formation, there are mainly two types of processes and their interactions are considered: the morpho-dynamic

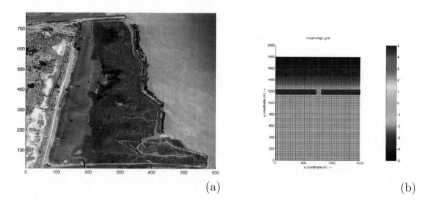

(a)   (b)

Figure 6.1: Model domain, bathymetry datasets and the coastline. (a) The prototype of a tidal salt marsh. The channels are formed after the sea side bank was open; (b) Grid and initial bathymetry for the schematized salt marsh model.

processes and ecological processes. The morphodynamic processes include: tidal flow, tidal asymmetry (residual current), suspended sediment transport, bed load transport (with bed slope effect), adjacent dry cell erosion and avalanching, sediment sorting in vertical and horizontal direction; mud segregation, and bed level update. The tidal flow is regarded as the driving force of the channel formation. Suspended sediment transport is responsible for the sediment exchange between the offshore area and tidal salt marsh (sediment net flux), the exchange between the shoals and channels within the tidal marsh, and adjacent dry cell erosion and avalanching are linked to bank erosion and channel formation. Ecological processes include vegetation growth and mortality, spatial expansion, seed dispersal and competition between themselves. The mortality process may be determined by: 1) Temperature cycle; 2) Lifespan; 3) Ambient force (flow shear stress, inundation time and depth etc.); 4) Competition. The competitions between two species are not included in this schematized model. Lifespan for Spartina may be long if the living conditions are favorite. So it is excluded in this model as well. The initial distribution of the vegetation is 50% everywhere within the tidal basin. The initial height and age are randomly distributed between 0 to 0.25 of the maximum height and age. The temperature in a year is shown in Figure 6.2. This annual temperature cycle is imposed repeatedly for 30 years in this simulation.

The processes for vegetation-flow interaction are the following two: 1) extra drag force for flow induced by vegetation stems and leaves; 2) enhanced turbulence generated by flow around the stems. The processes responsible for the vegetation effects on suspended sediment transport are the following two: 1) the vegetated bed reduces the entrainment of sediment into suspension (lower source); 2) more streamwise momentum is absorbed by the plants via drag forces (lower sediment transport capacity) (Lopez and Garcia, 1998). Vegetation root systems (tussock) vary seasonally and annually, binding sediment grains and rendering the sedimentary surface more resistant to erosive force (lower source), and hence reduce bed load transport at certain bed level.

Table 6.1: Parameter lists and morphodynamic model settings

Module	Parameter	Value	Description
Flow	Δt	1 [min]	Computational time step
	C	30 $[m^{0.5}/s]$	Chezy, bed roughness coefficient
	υ_H	1 $[m^2/s]$	Horizontal eddy viscosity
	D_H	10 $[m^2/s]$	Horizontal eddy diffusivity
Sediment	D_{50}	0.120 [mm]	Single fraction, non-cohesive
	ρ_s	2650 $[kg/m^3]$	Sediment density
	d	50 [m]	Initial erodable sediment depth at bed
Morphology	$morfac$	90 (-)	Morphological factor
	ThetSD	1	Factor for erosion of adjacent dry cells
	T	4 $[month]$	Hydrodynamic simulation time period
	T$_{mor}$	4 $month\times$ 90 $\simeq$ 30 $[years]$	Morphological simulation time period
	$\Delta t_{ecostep}$	120 $mins\times$ 90 $\simeq$ 7.5 $[days]$	Ecological simulation time step

6.2.3   Results

Vegetation dynamics

Figure 6.2 shows time series of ambient temperature and the corresponding vegetation population density. The mortality process is determined by the ambient temperature cycle and competition within the species in this case. The vegetation density follows the logistic growth rule and is interrupted by the temperature cycle every year. The representative heights and diameters of the vegetation increase linearly to the maximum values and decrease rapidly when the population density drops down.

Figure 6.3 shows snapshots of the typical spatial distribution of vegetation on the tidal flat during summer and winter seasons. Patch (tussock) patterns of vegetation are visible. The model starts with vegetation distributed homogenously in the whole domain. The patch pattern in Figure 6.3a, b clearly demonstrates the seasonal effects on the vegetation population dynamics. It also implies that the corresponding time scales of the relevant ecological processes and morphological processes are significant. For instance, during summer time, it is possible that ecological processes might be dominant if morphological force is not significant (i.e. not during storm period).

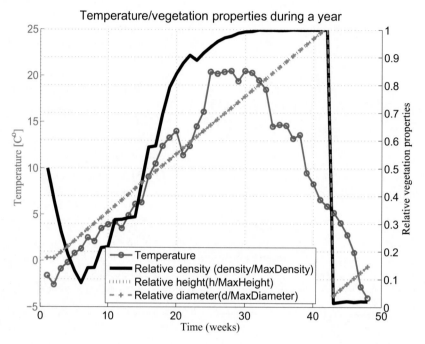

Figure 6.2: Varying ambient temperature and relative vegetation properties during a year

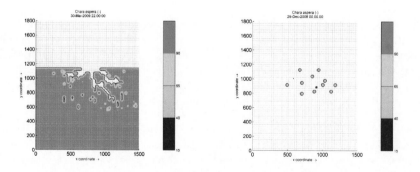

Figure 6.3: Snapshot of the typical spatial patterns of vegetation distribution (a) during the summer seasons and (b) during the winter seasons.

Table 6.2: Parameter lists and ecological model settings for Spartina *spp.*-like species

Module	Parameter	Value
Spartina (like)	Max density	1000 $[stem/m^2]$
	Max height	2 $[m]$
	Max Diameter	0.01 $[m]$
	Max Lifespan	12 $[years]$
	Growth rate	0.3 $[1/week]$ (if temperature $> 12^o$)
	Growth rate	0.5 $[1/week]$ (if $0^o <$ temperature $< 12^o$ and tempdiff$> 1^o$)
	Growth rate	-0.2 $[1/week]$ (if $6^o <$ temperature $< 12^o$ and tempdiff$< 1^o$)
	Growth rate	-0.5 $[1/week]$ (if temperature $< 0^o$)
	Inundation depth	1 $[m]$
	Salinty	30 $[ppt]$
	Shear stress	0.2 $[Pa]$

Morphological evolution

Without ecological effects, the morphological evolution of 30 years is shown in Figure 6.4a, while Figure 6.4b shows the morphological evolution with vegetation effects coupled once every week and Figure 6.4c shows the morphological evolution with vegetation effects coupled once every year. In general, evolution of morphological patterns is well recognized. The ebb and flood tidal delta are formed, and the banks are eroded and the tidal inlets are widening in all cases. Without vegetation effects (Figure 6.4a), the flood tidal delta attached with subchannel systems are quite symmetric, and the tidal inlet is wider than the cases with vegetation effects. With high frequent coupling with vegetation effects (Figure 6.4b), the widening of tidal inlet is narrower than other two cases, which implies that more energy in the system has been dissipated by vegetation. The flood tidal delta is highly asymmetric and much smaller in volume. There are much fewer sub-channels on the delta compared to the morphological changes without vegetation effects. However, with low-frequent coupling with vegetation effects (Figure 6.4c), the flood tidal delta is less asymmetric. The volume and range of the flood tidal delta are more close to the case without vegetation. There are fewer sub-channels on the delta compared to the morphological changes without vegetation effects. Once a patch of vegetation survives from the ambient force and starts to develop, channels attract more flow, which increases the possibility of vegetation to grow on the higher area.

Figure 6.6 shows the snapshot of the morphological change every 5 years till 30 years without vegetation effects. Initially the tidal inlet is quite narrow. After 5 years, the width of the tidal inlet increases significantly and the shape of ebb delta is clearly visible. The flood channel and delta inside the basin are remarkable. With times, symmetric tidal channels and deltas are formed. The tidal inlet width tends towards its equilibrium, which is confirmed in Figure 6.9 as well. The last figure in Figure 6.6 shows the cumulative

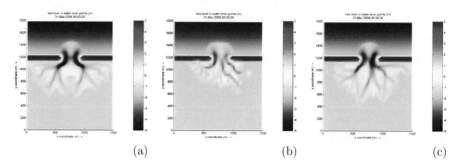

Figure 6.4: Morphological developement after 30 years (a) without vegetation; (b) with vegetation, where the information of vegetation and morphodynamics are exchanged (coupled) once every week, starting at winter (refer to Figure 6.8); (c) with vegetation, where the information of vegetation and morphodynamics are exchanged (coupled) once every year.

sedimentation and erosion depth.

Figure 6.5 shows a time series of vegetation percentage at the basin center (M=25, N=30, refer to Figure 6.1 for the grid arrangement) and the corresponding bed shear stress, total water depth and temperature at the moment shown in Figure 6.7 and Figure 6.8. The green boxes indicate every five morphological years (every 20 hydrodynamic days) when the snapshot of vegetation distribution and morphology patterns are taken at Figure 6.7 and Figure 6.8. From the time series of vegetation percentage at the basin center, we see that the initial condition for vegetation is not crucial as well. The percentage of vegetation occupation converges for both models starting from summer and winter.

Basin characteristics

Figure 6.9 demonstrates morphological characteristics of the tidal inlet with and without ecological effects. The significance of the coupling time step between ecological processes and morphological processes is also clearly demonstrated. The scenario for coupling ecological information once every year (the green lines with marker + in the Figure 6.9) gives out much less effects than the scenario for coupling once every week (the blue dash lines) on the morphological changes. Figure 6.9a indicates the cross section area of the inlet changing with time. The initial width is rather narrow. As time goes by, the cross section area of the inlet increases. After 30 years of development, the morphology of this tidal inlet system tends to get stable. With ecological effects (the scenario for coupling once a week), the cross section area of the tidal inlet is around 30% less than that without ecological effects after 30 years.

Figure 6.9b indicates the volume of sediment imports into or exports towards the outer sea with time. Initially there is no ebb delta. After 30 years of morphological evolution, an ebb delta is formed and the bank and bottom of the inlet are eroded. So the system is

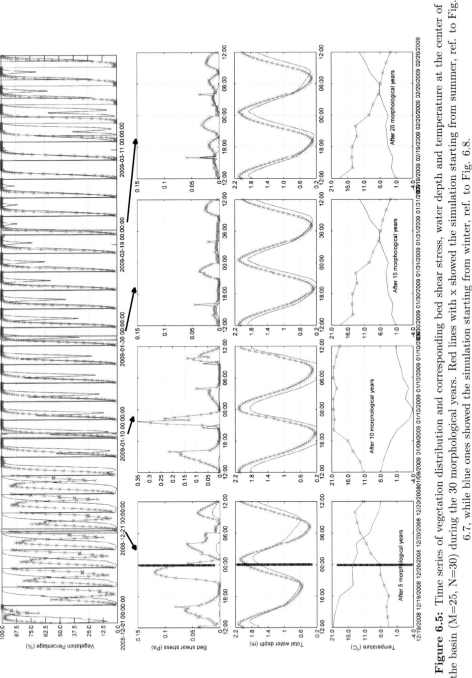

Figure 6.5: Time series of vegetation distribution and corresponding bed shear stress, water depth and temperature at the center of the basin (M=25, N=30) during the 30 morphological years. Red lines with x showed the simulation starting from summer, ref. to Fig. 6.7, while blue ones showed the simulation starting from winter, ref. to Fig. 6.8.

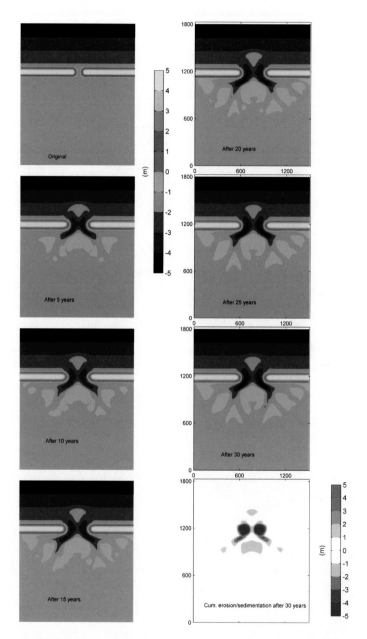

Figure 6.6: Bedlevel change and cumunative erosion and sedimentation during 30 years. No vegetation effects.

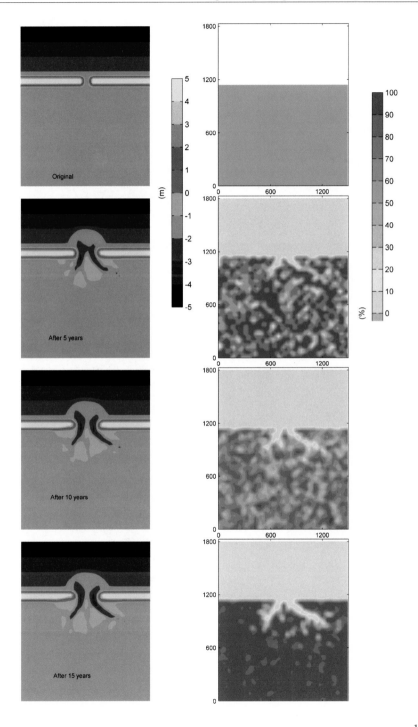

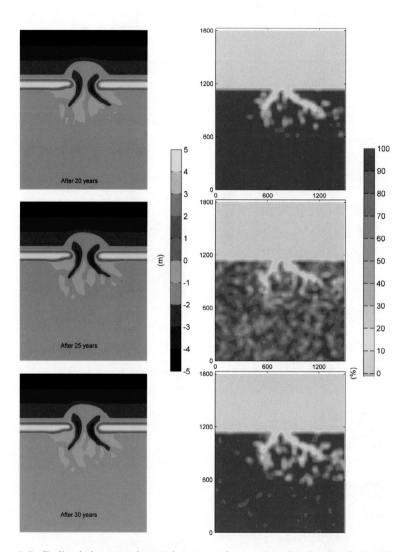

Figure 6.7: Bedlevel change and spatial pattern of vegetation distribution during 30 years. Coupled every week, starting from summer (refer to the red x lines in Figure 6.6).

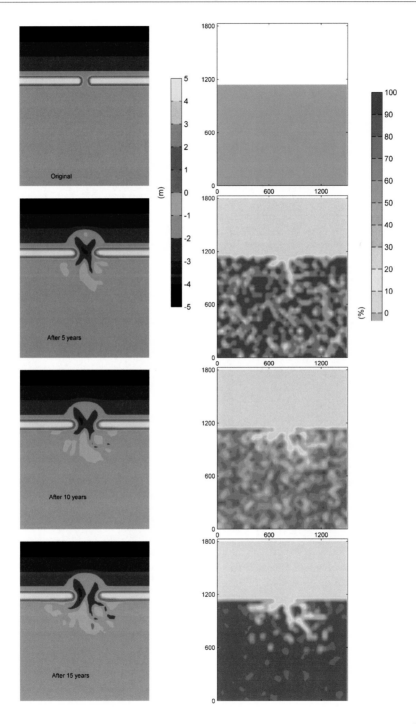

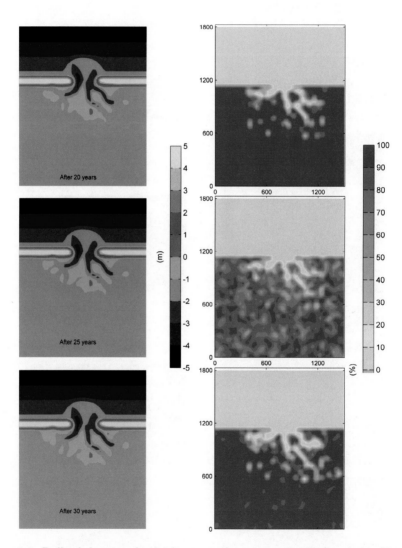

Figure 6.8: Bedlevel change and spatial pattern of vegetation distribution during 30 years. Coupled every week, starting from winter (refer to the blue lines in Figure 6.6).

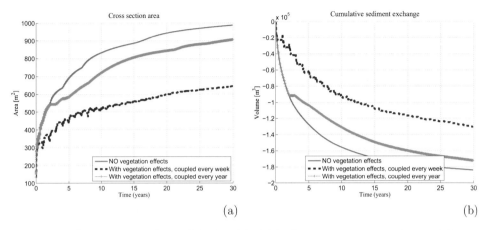

(a)   (b)

Figure 6.9: Charactiristics of the tidal inlet. (a) Tidal inlet cross section area change with time; (b) Sediment exchange through the tidal inlet (+ import to the basin / - export to outside of the basin).

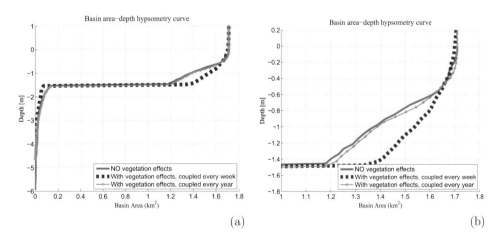

(a)   (b)

Figure 6.10: Basin area-depth hypsometric curve after 30 years morphological development. a) The whole basin; (b) Zoomed in the intertidal area.

133

exporting sediment towards the out sea. Nevertheless, with ecological effects coupled once every week, the export volume is significantly (50%) smaller than that without ecological effects and the scenario with ecological effect coupled once every year. First of all, it implies that with vegetation, much less sediment has been exported from the basin out towards the sea. This might provide an option for management of the barrier island type of coast: to have vegetation field, in form of salt marsh, etc. might be able to enhance sediment deposition to adapt to the sea level rise, and dissipate hydraulic loads from the sea and be beneficial to coastal defenses. Secondly, it demonstrates the significance of the coupling time step between ecological processes and morphological processes. The volumes of the ebb delta are almost identical between the scenario without ecological effects and the one coupled once every year. Probably the frequency (of coupling once every year) is not sufficiently high to represent the cycle of the natural system.

Figure 6.10 indicates the basin area-depth hypsometric curve after 30 years of morphological development. For the whole basin, the maximum depth is around 6m, located at the inlet, and the surface below the mean water level is around 1.5 km^2. In general, for the scenario with ecological effects coupled once every week, there are larger numbers of deep and narrow channels in the basin area than the results from the scenario without ecological effects. It also demonstrates the significance of the coupling time step between ecological processes and morphological processes clearly. When coupled once every week, a large number of narrow channels appear in the basin, while coupled once every year (similar to the case without vegetation effects), a few wider, shallower channels appear in the basin.

6.2.4   Conclusions

In this section the new bio-geomorphological modules are applied to simulate the effect of vegetation on the morphodynamics of a schematized tidal inlet similar to those studied by Roelvink (2006); Tung et al. (2009); Dissanayake et al. (2009). The analysis focuses on the effects of vegetation on the channel network development in the tidal basin and the overall characteristics of the tidal basin. However, even though the model is highly schematized with minor calibration efforts, the results still show the importance of considering the effects of ecological processes. Furthermore, the significance of the coupling time step between ecological processes and morphological processes is demonstrated. More frequent coupling can resolve the tidal forces. For ecological systems which are usually with strong nonlinear characteristics, ambient forces at different coupling moments might lead to dramatic differences in the system evolution. The extended integrated generic geomorphological model is a tool with high potential to reveal the significance of vegetation effects on morphological development.

6.3 Case 2: Bio-geomorphological modelling in

Nisqually estuary, Puget Sound

6.3.1 Introduction

The wetland in Nisqually estuary, south of Puget Sound, US, is planned to be restored by removing the dike ring along the seaside (Gelfenbaum et al., 2006; Puget Sound Partnership, 2008; Nisqually River Basin Plan, 2008; US Fish and Wildlife Service, 2005). Concerns have been raised on the ecological features and morphology change in the near future caused by the dike removal project (Monden, 2010). Three geographical elements are involved, i.e. Puget Sound (salt water, hardly any sediment), Nisqually River (fresh water and sediment source) and the Nisqually estuary between them. The main hydrodynamic forces then include tidal force and river flow. Sediment types are varying from sand to clay. More than 12 vegetation species exist in this area. The complex interaction of ecological processes and morphological processes make this area of grate interest for integrated eco-morphological modelling practice. The project groups from USGS and Deltares are gratefully acknowledged for providing the data for this application.

Puget Sound

Puget Sound is located between the Cascade and Olympic mountains, in the state of Washington in the northwest of the United States (Finlayson, 2005). Formed by glaciations from 10,000 years ago, it is characterized as a system of deep and narrow channels divided by islands and peninsulas. The area is approximately 160 by 80 kilometers and consists of 4 deep basins separated by shallow sills. The average depth is 62 m and the maximum depth, just north of Seattle, is 280 m. Puget Sound is connected with the Pacific Ocean through the Strait of Juan de Fuca. Tides in Puget Sound are of the mixed type, with two high and two low tides per day. The configuration of basins and sills causes the tidal range to increase when propagating through the sound. Large tidal influence combined with the shallow sills and narrow passages leads locally to large velocities, up until 9 or 10 knots at Deception Pass. This also affects the sediment transport in the estuary, and makes that only a small amount of sediment can leave basin area. Thus in this case, we assume that the Puget Sound side is not the main sediment source for Nisqually estuary and the salt marsh area.

Nisqually River

The Nisqually River drains the southern slope of Mount Rainier, part of the Cascade mountain range (Burg & Tripp, 1980). It flows approximately 130 km, west-northwest, to the Nisqually Estuary where it flows into the southern end of Puget Sound, near Olympia. The watershed area is approximately 1970 km^2. The Nisqually River has an average discharge of approximately 60 m^3/s in winter and 30 m^3/s in summer. Peak discharges can be as high as 600 m^3/s for a 10 year return period (Puget Sound Partnership 2008). There is no data available for sediment transport through the river into the estuary. In the model, sediment discharge data from an adjacent river is used as the analogy.

Nisqually estuary and restoration projects

Nisqually estuary is located on the border of Thurston and Pierce Counties, 16 km northeast of Olympia, Washington. Several restoration projects have been carried out since 2002. In the east bank of the river, tidal inundation has been restored to an area of approximately 16 hectares in 2002 and an additional 40 hectares in 2006. In the west bank of the river, a larger restoration project has been carried out in 2009. The old Brown Farm Dike enclosed a rectangular area of approximately 2.5 by 2 kilometers were removed. On the north it is bordered by Puget Sound, on the west by a small stream called McAllister Creek and on the south by the Highway I-5.

The restoration projects

The situation before the restoration project is shown in Figure 6.11. Even after a century of being shut off from tidal influence the old channels within the dike rings were still visible before the restoration. Though they are partly covered with vegetation, four major channels can be distinguished: three along the northern dike and one along the western dike next to McAllister Creek. The majority of the area was covered with grasslands, with some riparian and mixed forest habitats in the slightly higher eastern part. Along the western dike was a part with lower elevation which was submerged most of the time. Over 50 % of the area was covered with the tall reed canary grass (Phalaris arundinacea, fresh water species), an invasive species that suppresses native vegetation and reduces biological diversity (Woo et al., 2010).

Outside of the dike some patches of salt marsh remained. The tidal channels were still in line with the channel remains inside the dike, which suggests that the system was more or less in equilibrium. Some of the dominant salt marsh vegetation species are the Seashore Saltgrass (*Distichlis Spicata*), Pickleweed (*Salicornia Virginica*), Tufted hair-grass (*Deschampsia Cespitosa*), Perennial Ryegrass (*Lolium Perenne*) and Lyngbyes sedge (*Carex Lyngbyei*) (Burg & Tripp, 1980; Woo et al., 2010).

The plans for the restoration project include two steps. Firstly a new dike was constructed to protect a small part of the area in the southeast, which contains a visitor centre and some other buildings. Secondly, after this dike was completed the original outer dike was removed, exposing an area of 308 hactares to the tides of Puget Sound. The effect of this became visible very soon (Figure 6.11). Almost all of the freshwater vegetation died off, including all of the invasive reed canary grass. The old channels got inundated again and most of the vegetation got flushed out. The channels also seem to be expanding further into the area. All of this makes it clear that the area is in a state of transition at the moment, and that it will keep changing considerably in the coming years.

 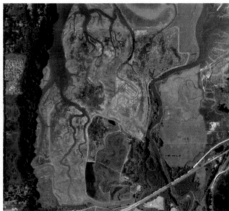

Figure 6.11: Nisqually estuary before (Nov., 2009 before the removal of the dike, in the left panel) and 7 months after dike removal (June, 2010, in the right panel) (from Google Earth)

6.3.2 Model setup and dominant processes

Hydrodynamic model setting is briefly listed in Table 6.3. It is believed that main sediment source for the morphological changes are the river input in this area. Unfortunately, there is no measurement for both flow and sediment on Nisqually River. Two hydrodynamic regimes are modeled and the morphological change for 6 years and the vegetation distributions of various types are examined. The high flow regime has 3 times higher input of river flow and thus sediment influx than the low flow regime. The morphodynamic processes include: tidal flow, tidal asymmetry (residual current), suspended sediment transport, bed load transport (with bed slope effect), adjacent dry cell erosion and avalanching, and bed level update. The tidal flow and river discharge are the main forces. Suspended sediment transport is responsible for the sediment exchange between the offshore area and tidal salt marsh (sediment net flux), the exchange between the shoals and channels within the tidal marsh, and adjacent dry cell erosion and avalanching are linked

(a)   (b)   (c)   (d)

Figure 6.12: Vegetation species modeled in this case. a) Phar (*Phalaris Arundinacea*), a type of initial fresh water vegetation; b) Caly (*Carex Lyngbyei*) as a type of pioneer salt marsh vegetation; c) Dist (*Distichlis Spicata*) , a type of salt marsh vegetation; d) Savi(*Salicornia Virginica*), a kind of slow spreading salt marsh vegetation (pictures from http://plants.usda.gov/).

to bank erosion and channel formation. Ecological processes include vegetation growth and mortality, spatial expansion, seed dispersal and competition within one species and between two species. In this modeling practice, 4 species are taken into account (Figure 6.12). They are (from left to right) : Phar (*Phalaris Arundinacea*), a type of initial fresh water vegetation; Caly (*Carex Lyngbyei*) as a type of pioneer salt marsh vegetation; Dist (*Distichlis Spicata*) , a type of salt marsh vegetation; Savi(*Salicornia Virginica*), a kind of slow spreading salt marsh vegetation. The interactions between species are represented as competetion for the limited resource in the segments. The initial distribution of each species are sketched based on field observation (Monden, 2010). The ecological model setting is briefly listed in Table 6.4. The processes for vegetation-flow interaction, vegetation effects on sediment transport are similar to the descriptions in Section 6.2.2. The model domain is shown in Figure 6.13a. Four polygons are defined in the model domain. Polygon 1 covers the restoration area, Polygon 2 shows the river area, while Polygon 3 shows the original marsh area and Polygon 4 represents the rest area within the model domain. Figure 6.13 b shows the initial spatial distribution of the vegetation species in the model domain.

In this case, the vegetation species are regarded to have multi-year reproductive cycles. Therefore the mortality processes are less important than the competetion process and the temperature (seasonal) control is not activated in this case. The competetion rules and the spatial expansion rules are described in Table 6.5.

Table 6.3: Parameter lists and morphodynamic model settings

Module	Parameter	Value	Description
Flow	Δt	1 [min]	Computational time step
	Grid size	30 [m] * 30 [m]	-
	B.C. P. Sound	Tidal components	Nested from a larger model
	B.C. River	Time series discharge	Estimated
Sediment	Sand D_{50}	0.128 [mm]	Initial thickness 1 [m]
	Silt	$w_s = 0.25$ [m/s]	Initial thickness 4 [m]
	Clay	$w_s = 0.0046$ [m/s]	Initial thickness 1 [m]
	B.C. P. Sound	0	Estimated
	B.C. River	conc. time series	Estimated
Morphology	*morfac*	72 (-)	Morphological factor
	ThetSD	1	Erosion factor of adjacent dry cells
	T	1 [*month*]	Hydro. simulation time period
	T$_{mor}$	$1\ m \times 72 \simeq 6$ [*years*]	Morpho. simulation time period
	$\Delta t_{ecostep}$	$2\ hours \times 72 \simeq 6$ [*days*]	Ecological simulation time step

6.3.3   Results

Because the sediment from Nisqually River is supposed to be the main sediment source on the tidal flat, and also the discharge of river has strong seasonal variation, the river water discharge and sediment input setting may be essential to the morphology development. For lack of data, two regimes are defined in this application: the high flow regime and the low flow regime. The river discharge and thus the overall sediment input from the river to the system in the high flow regime thus are 3 times larger than the settings in the low flow regime.

Low flow regime

Without vegetation, most of the sediment, mainly fine sediment, from the river, settled in the river and the area close the river (Figure 6.14a). The sedimentation volume in the original embanked area is around 80k m^3 without counting the effect of vegetation dynamics.

With vegetation, the active processes are then tidal current, sediment transport and morphology change, vegetation growth and spreading side by side. Because this area

Table 6.4: Parameter lists and ecological model settings

Parameter	Values	
	Phar	*Caly*
Max density	600 $[stem/m^2]$	600 $[stem/m^2]$
Max height	1.18 $[m]$	0.3 $[m]$
Max Diameter	0.01$[m]$	0.01$[m]$
Growth rate	0.7 [1/week](at 15°)	0.5 [1/week](at 15°)
Inundation depth	1.1 $[m]$	1.44 $[m]$
Salinty	3 $[ppt]$	20 $[ppt]$
Shear stress	0.26 $[Pa]$	0.26 $[Pa]$
	Dist	*Savi*
Max density	1000 $[stem/m^2]$	800 $[stem/m^2]$
Max height	0.34 $[m]$	0.30 $[m]$
Max Diameter	0.01$[m]$	0.01$[m]$
Growth rate	0.3 [1/week](at 15°)	0.1 [1/week](at 15°)
Inundation depth	0.92 $[m]$	1.18 $[m]$
Salinty	31 $[ppt]$	31 $[ppt]$
Shear stress	0.26 $[Pa]$	0.26 $[Pa]$

is quite well sheltered, the wave effect was not taken into account. With effects of vegetation dynamics, the sedimentation in the river decreases and more sedimentation happens on the tidal flat area where the vegetation grows and spreads (Figure 6.14b). The sedimentation volume in the original embanked volume increased 10k m^3 higher, which in turn helps the vegetation spread because of higher elevation after sedimentation.

Vegetation distribution is also represented reasonably well (Figure 6.15a and 6.15b). At the beginning, the fresh water species *Phar* retreats significantly. And gradually, the pioneer species, *Caly*, spreads and occupies most of the space. Competition of the 3 types of salt marsh species is also observed. The slow growing species, *Dist* and *Savi* stays within the area where it has stayed already for 100 years, outside the bank.

High flow regime

It is hypothesized that the sediment from river is the main sediment source on the tidal flat, and since also the discharge of river has a strong seasonal variation, the river water discharge and sediment input setting may be essential to the morphology development. In this regime, the river discharge and thus the overall sediment input from the river to the system thus are increased 3 times compared to that in the low flow regime. Significant effects are observed. First of all, all the river areas are generally under erosion, and the sedimentation areas are significantly wider (Figure 6.14c and Figure 6.14d). The sedimentation volume is also much higher because of higher input. Second, the vegetation

Table 6.5: Rules of competition & spatial expansion

Rules				
	Phar	*Caly*	*Dist*	*Savi*
Too crowded	50 % growth	10 % growth	10 % growth	10 % growth
τ_{cr} restriction	70 % left	70 % left	70 % left	70 % left
Salinity restriction	10 % left	10 % left	10 % left	10 % left
Lateral expansion				
If 2 over 8 neighbouring cells are higher than 50 %, or 1 cell higher than 75 %	+20 stems per week	+15 stems per week	+15 stems per week	+15 stems per week

dynamics are also very different. The fresh water species, *Phar* retreated much less with lower salinity in the domain. And the pioneer species, *Caly* also spread faster than that in low flow regime. The fast spread of Phar and Caly also suppressed other species, *Dist* and *Savi* to grow and to extend as well (Figure 6.15c and 6.15d).

6.3.4 Conclusions

In this case the morphological change within a mid-term period (6 years) after the outer bank is opened is qualitatively modeled. Even though the parameters in the ecological processes are still far from perfect, the vegetation population dynamic processes are implemented in the model, and spatial patterns of vegetation distribution are reproduced qualitatively. Interactions of physical processes and ecological processes are represented reasonably well. The effects of vegetation processes on the morphological development are significant. With vegetation effect, sedimentation in the river decreases and sedimentation on the tidal flat area increases where the vegetation grows and spreads. The sedimentation volume in the original embanked area is up to around 90k m^3 with vegetation effects. The sedimentation volume with vegetation effects is higher than the volume without vegetation, which in turn helps the vegetation spread because of the higher elevation after sedimentation. Also for the existing salt marsh area, after the vegetation has been built up, the erosion volume in that area is much less than that without vegetation.

Due to the limitation of field flow and sediment data from the Nisqually river, the model is not validated quantitatively. Continuous monitoring of morphological change of this area will be helpful to calibrate the morphological model. Furthermore, even though the physical processes described in the Delft3D system are quite sophisticated, to be a robust, wide applicable modeling tool for bio-geomorphodynamics, there are still spaces to improve the model. Mechanisms of interactions between physical processes and ecological processes, and also the ecological dynamics are essential subjects for further research and collaborations.

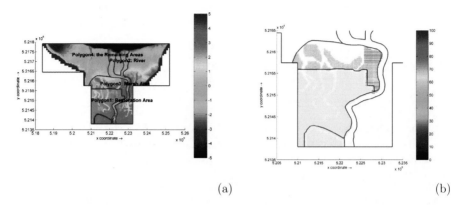

(a)   (b)

Figure 6.13: Model domain. (a) Initial bathmetry, grid arrangement and morphological units (4 units are represented into 4 polygons); (b) Initial distribution of vegetation species, the color of the markers shows the percentage of capacity for each species. The species Phar (*Phalaris Arundinacea*) showed as o; Caly (*Carex Lyngbyei*) in +; Dist (*Distichlis Spicata*) in x, and Savi (*Salicornia Virginica*) in diamond.

6.4   Case 3: Shoreface Connected Radial Sand ridge

at southern Yellow Sea, China

In this section, an unique large morphological feature, shoreface connected radial shape sand ridges located at southern Yellow Sea, China is described.

This area is famous for its both highly dynamic morphology and ecology. From around 1128 AD, the Yellow River changed its main course and discharged into the sea at this area. A huge amount of fine sediment was brought from upstream and settled in the coastal area. The abundant sediment supply in couple with the strong and stable tidal flow pattern put the area in high dynamic situation. Till 1855 AD, the Yellow River moved back to the north course. From then on, the sand ridges area loses the main sediment supply source, and has been under erosion in general. The abandoned Yellow River Delta was flattened and eroded underwater gradually and coastlines continued retreating. Within this morphologically dynamic area, ecology is also highly dynamic (see e.g. Ke et al. (2009), and Zuo et al. (2009)). The immense tidal flat areas are important stopover sites for migration bird species. Within the rich amount of intertidal habitat, some of the areas are exploited as aquafarms for seafood. Since 1979 *Spartina alterniflora* is introduced from Florida to this area to enhance sedimentation along the coast. Figure 6.16a showed the distribution of *S. alterniflora* salt marsh on Jiangsu coast in 2001 (Zhang-RS et al., 2004).

This area is also famous for the heavy anthropogenic impact. The area was formed in

142

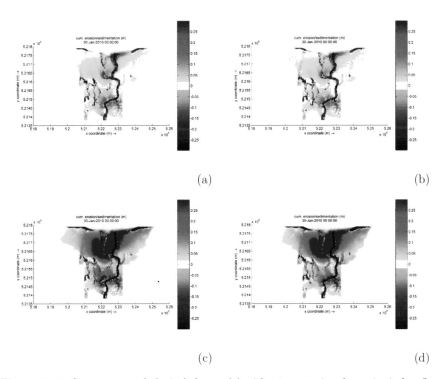

(a)

(b)

(c)

(d)

Figure 6.14: Six-years morphological change (a) without vegetation dynamics in low flow regime , and (b) with vegetation dynamics in low flow regime; (c) without vegetation dynamics in high flow regime , and (d) with vegetation dynamics in high flow regime;

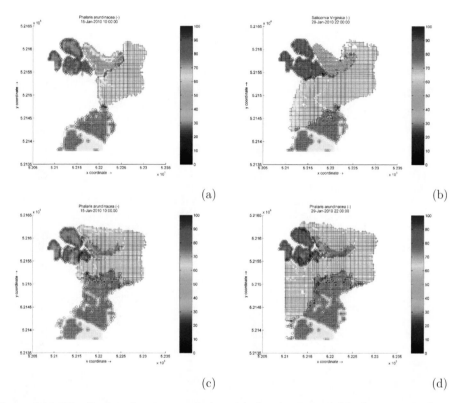

(a)   (b)

(c)   (d)

Figure 6.15: Distribution of various vegetations (a) after 3 years and (b) after 6 years in low flow regime; Distribution of various vegetations (c) after 3 years and (d) after 6 years in high flow regime;. The color of the markers shows the percentage of capacity for each species. The species Phar (*Phalaris Arundinacea*) showed as o; Caly (*Carex Lyngbyei*) in +; Dist (*Distichlis Spicata*) in x, and Savi (*Salicornia Virginica*) in diamond, (From the model!)

recent 1000 years and has been heavily modified by people lived in this area. The deep channels between the tidal sand ridges have been utilized for navigation for centuries. In the recent decade, part of tidal flat is exploited for land development to relieve the increasing population stress. Figure 6.16b showed the reclamation plan before 2020 of the tidal flat in Jiangsu coast (Zhang-C.K. & Wang, 2009). It is also shown the detailed reclamation on the tidal flat in Figure 6.17a and the red crowned crane national nature reserve in Figure 6.17b, which also shown the requirement to get better knowledge for this area for comprehensive utilizations.

However, the understanding of the water-sand-ecology system is still quite limited. To better understand this area, many research topics are to be carried out. One of the most important topics is then the generation and long-term evolution trend, which are focused on in this study. The generic geomorphological model developed in the previous chapters is potentially a suitable tool. Nevertheless, the power of the generic geomorphological model lies in its flexibility for unstructured grid and extendibity towards biomorphological model. These features are not really highlighted at present stage of this study. Thus hereafter the open source version of present Delft3D online MOR is applied in the modeling practice. Nevertheless, the generic geomorphological model still shows its potential and will be applied in the future research when the biomorphological processes are focused on in this area. Results from a first experiment show that even though morphological processes are rather complex, the generic geomorphological model still provides rather similar morphological patterns compared to the results from the Delft3D online MOR (Figure 6.18).

6.4.1 Introduction

Shoreface connected sand ridges are large scale morphological features. They are widely found along the continental shelves. Examples are the shelves along the east coast of the United States (Swift et al, 1978), Germany (Anita, 1997), Netherlands (Van de meene and Van Rijn, 2000), Argentina (Parker et al, 1982) and China (Yan et al, 1999a, b). Different from the offshore sandbanks, the shoreface connected sand ridges are characterized by their elongated rhythmic bed forms in scale of tens to hundreds of kilometers, extended from coast line to 20m in water depth, from offshore end of the shoreface to the beginning of the outer shelf. Unlike other the shoreface connected sand ridges in planar parallel form, the sand ridges at southern Yellow Sea are distinct for their radial shape. The spatial scale is about hundreds of kilometers and time scale is limited within the recent 150 years. It fans out with a central angle of about 160 degrees in the N, NE, E and SE directions, with Jianggang as the radial center. Hereby, a brief review is given on the study on large scale shoreface connected sand ridges.

In addition to field monitoring and in-site survey, processes based modeling has proven advantages to help understanding natural phenomena at large scales. Two approaches have been applied to study shoreface connected sand ridges:

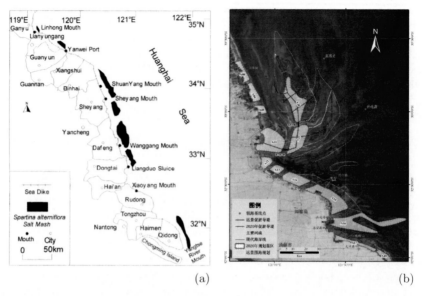

(a)

(b)

Figure 6.16: (a): The distribution of S. alterniflora salt marsh on Jiangsu coast in 2001 (from Zhang-RS et al, 2004); (b): The reclamation plan of the tidal flat in Jiangsu coast (from Zhang et al, 2009). The red polygons are the reclamation area and the red polylines indicates the groins to stabilize the large tidal flat area. The reclamation areas are numbered in A08 to A16.

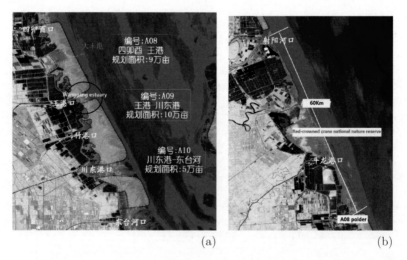

(a)

(b)

Figure 6.17: (a): An example of the reclamation area A08 on the tidal flat; (b): The red crowned crane national nature reserve located at north of A08 (Both from Zhang et al, 2009).

146

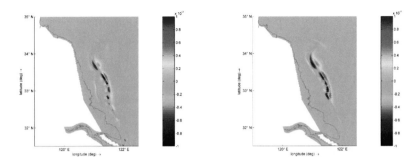

Figure 6.18: Cumulated sedimentation/erosion after one tidal cycle, with morfac as 1. The results from Delft3D online MOR are shown in the left panel and the results from the generic geomorphological model described in the previous chapters are shown in the right panel. Erosion is indicated in blue and sedimentation in red.

i) Full process-based modeling, including non-linear processes (Zhang and Zhang, 1996, 1998; Yan et al, 1999a, b; Zhu et al, 1995; Zhu, 2003; Chen, 2008);

ii) Idealized process-based modeling, such as stability analysis (Trowbridge, 1995; Falques et al, 1998; Calvete et al, 2001a, b, 2002, 2003; Walgreen et al, 2002; Swart et al, 2003, 2008).

Shoreface connected sand ridges are widely found along the continental shelves. Examples are the shelves along the east coast of the United States (Swift et al., 1978), Germany (Antia, 1996), Netherlands (Van De Meene & Van Rijn, 2000), Argentina (Parker et al., 1982) and China (Yan, Zhu, & Xue, 1999),(Yan, Song, et al., 1999). Different from the offshore sandbanks, the shoreface connected sand ridges are characterized by their elongated rhythmic bed forms on scales of tens to hundreds of kilometers, extended from coast line to 20m in water depth, from offshore end of the shoreface to the beginning of the outer shelf. However, unlike other shoreface connected sand ridges of planar parallel form, the sand ridges at southern Yellow Sea are distinct for their radial shape. They fan out with a central angle of about 160 degrees in the N, NE, E and SE directions, with Jianggang as the radial center (Figure 6.19).

The area is both ecologically and economically important. The deep channels between the ridges have been utilized for navigation for centuries. The immense tidal flat areas are important stopover sites for migration bird species. Within the rich amount of intertidal habitat, some of the areas are exploited as aqua farms for seafood. Part of the tidal flats are exploited for estate development to relieve the increasing population pressure. Nevertheless, the understanding of the water-sand system is still quite limited. Geological analysis on drilling data sets in the radial sand ridges area shows that the sediment supply includes both fluvial sediments from adjacent river systems and marine sediments by underwater erosion at shoreface/outer shelf and redistribution Zhang-RS and Chen (1992). Hydrodynamic simulation show a radial tidal wave pattern at large spatial scale

147

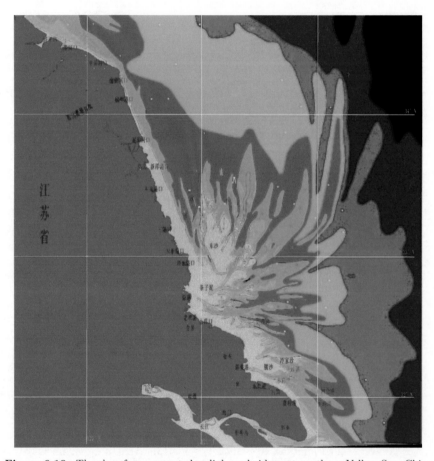

Figure 6.19: The shoreface connected radial sand ridges at southern Yellow Sea, China (sketch map based on bathymetry of 1998)

(Ren, 1986; Zhang & Zhang, 1996), which leads to a hypothesis that the radial sand ridges are the morphological response to the radial wave pattern. All those studies assumed that strong tidal currents are dominant in generation and maintenance of the sand ridges, due to the fact that the largest observed tidal range could be as high as 9 m, and strongest tidal currents could be 3 knots at certain channels during flood (Yan, Zhu, & Xue, 1999; Zhu, 2003; Lu, 2005; Chen, 2008). However, there is no general consensus on the mechanisms responsible for the initial generation in the past, maintenance at present and the future evolution of the sand ridges in radial shape. Moreover, the effect of waves has been regarded as one of the main driven forces to form shoreface connected sand ridges (e.g. Calvete, Falqués, et al. (2001)). It is not yet considered in previous studies explicitly. Therefore, it is necessary to carry out a morphodynamic study to improve the understanding of the radial morphological feature from a comprehensive point of view.

This study aims to reconstruct the radial spatial pattern of the sand ridges in the southern Yellow Sea, based on the historic bathymetry at 1855 AD when the Yellow River changed its route gradually northwards, and the sand ridges area lost its main sediment supply source from then on. In this study, a process-based numerical model is applied to investigate the formation and long-term development of this unique shoreface connected radial sand ridges. Tide current, wave, sediment transport and morphological evolution are simulated simultaneously (online coupled). It is assumed that tide and wave forcing complement to each other to form and to maintain the system at different scale.

The generic geomorphological model developed in the previous chapters was planned to be applied in the early stage of this study. Preliminary result show that it works out similar results as the results from the Delft3D online MOR. However, since this case is already very complex from a purely morphological point of view, we decided to carry out the morphodynamic analysis with the fast and robust present Delft3D online MOR, as a necessary first step towards fully coupled bio-geomorphological simulations.

6.4.2 Characteristics of the shoreface-connected radial sand ridges at southern Yellow Sea, China

Historic development and present planar form

The formation of the sand ridge system is divided into 4 stages: early Holocene; Mid Holocene to 1125 AD; 1128 to 1855 AD; after 1855 AD till now. During the early Holocene, the Yangtze River discharged into the sea and formed the estuarine delta at the area between Dongtai and Jianggang (Figure 6.19), shown by fine-grained sand layers found in the drilling cores (Wang-Y & Zhu, 1998). Throughout the mid Holocene to 1125 AD, the Yangtze River mouth moved southwards to the present position (the area of Shanghai city). The coastal area was under severe erosion due to lack of sediment supply. During the period of 1125 to 1855 AD, the Yellow River broke its bank, took over the course of Huai River and carried out huge amount of fine partials to the coast. The fine materials deposited at the estuary and formed the layers with relatively fine sediment (clay and silt) above the layers with coarse fine sand (fine sand), shown in the drilling cores (Wang-Y & Zhu, 1998). The abundant sediment supply, in couple with the strong and stable tidal flow pattern, is the prerequisite to form and to maintain the radial shape of the sand ridges. At the end of this period, the south and mid part of sand ridges were already in radial shape under water (Figure 6.21d). The last stage, from 1855 AD till now, is mainly focused in this study. After 1855 AD, the Yellow River changed its route gradually northwards to the location close to the present estuary, north to the Shandong Peninsula and discharged into the Gulf of Bohai. From then on, the sand ridges area loses the main sediment supply source, and has been under erosion in general. The abandoned Yellow River Delta was flattened and eroded underwater gradually and coastlines continued retreating. The -10m contour lines were 120 km away from coast line in 1904 AD and they were 20 km away from coast line in 1930, 12 km away in 1965, 8 km away measured in 1994, and only 2 km away from the coast line in 2004 (Chen, 2008). Part of the sediment went southwards to form the north wing of the shoreface connected sand ridges. Since 1855 AD, the sand ridges system has been under high dynamics to adapt to the tidal flow field and also affected the flow current significantly as feedback.

Today the remarkable huge sand body covers more than 20,000 km^2 area, located at the southern Yellow Sea, east of the coast in Jiangsu Province. More than 70 sand ridges are distributed from the mouth of the Yangtze River to the mouth of the Sheyang River extending around 200 km in north-south direction and around 90 km to the east. More than 10 of them emerge above the water during low tide period. The length of the sand ridges is from tens to near a hundred kilometers and its width is from several to tens kilometers. The big sand body has unique geomorphological feature with wide-spread ridges and troughs, which radiate in the N, NE, E and SE directions with Jianggang ($N32^o43.8'$, $E120^o50.7'$) as the radial center (Figure 6.19) (Zhang-RS & Chen, 1992; Yan, Zhu, & Xue, 1999; Yan, Song, et al., 1999; Zhu, 2003; Lu, 2005).

A number of large sand ridges and the deep channels in between are shown in Figure 6.19 (Zhu, 2003; Lu, 2005). In some channels the -10m bathymetry contour lines extend very close to the main land. From the sketched bathymetry contour lines over the past 40 years, the outer boundaries of the sand ridges area are under erosion, but the landside areas are in a high sedimentation environment (Chen, 2008).

Hydrodynamic characteristics

- **Temperature and salinity**

  This area belongs to monsoon climate zone. The yearly average temperature is 13-15C, while yearly participation varies from 900 mm to 1,100 mm. The north winds dominate during the winter, while south winds dominate during the summer. The maximum wind speed happens during the hurricane seasons between July to September (Chen, 2008).

Most of the rivers in this area are controlled by locks. The peak discharge happens during the flood season. One tenth of discharge from Yangtze River (yearly average $20,000m^3/s$) flowed northwards along the coastal, which desalinate the south flank of sand ridges area significantly during the summer (Wang-Y, 2002). At the She Yang River mouth (Figure 6.19), the average salinity could be around 30 during winter and 23 during summer (Zhu, 2003).

- **Tide wave and tidal current**

It is widely accepted that tide is one of the main forces in this area. Tidal circulation modeling shows that progressive wave from Pacific Ocean propagates from the southeast to this area and divides into two parts after passing by Ryukyu Islands (Zhang & Zhang, 1996). One part goes through the strait between Shandong Peninsula and Korean Peninsula northwards into Bohai Sea, while another part turns to be rotational tidal wave in anti-clockwise direction, due to Coriolis force and the reflection of the Shandong Peninsula. The two wave systems meet and superimpose in this area, inducing a radial tidal wave pattern and remarkably large tidal ranges. The mean tidal range is 6.4m at Jianggang, while the maximum tidal range increase to 8m, and the minimum tidal range can drop to 5m. The observed maximum tidal range can reach 9.28m (Zhu, 2003; Yan, Zhu, & Xue, 1999) measured outside of Xiao Yangkou Harbor (Figure 6.19). Analysis on the tidal wave energy rates shows that the tidal wave at north flanks, close to the amphidromic point of M2 (Figure 6.23c,d), have the characteristics of standing wave while in the south flank, it is characterized as the progressive wave, propagated from the outer sea (Yan, Zhu, & Xue, 1999; Lu, 2005). This is also verified by the regional continental shelf model (Bijlsma et al., 1998). The tidal ranges at various measurement stations along the coast in this area are shown in Figure 2.

Tidal currents in this area are dominated by semi-diurnal M2 constituent. The flow current in the channels, moving back and forth along the ridge lines, can be higher than 3 knots during the spring tide (Zhu, 2003). The tidal current ellipses point to Jianggang generally, which is also verified by the numerical simulation (Figure 7). The tidal asymmetry is significant. In many measurement stations at the south flank of the radial sand ridges, for example, South West Sun Sand, the ebb velocity is found much higher than flood velocity, and the ebb tide duration is 1 hour longer than the duration for flood during both spring and neap tide period, hence the ebb tidal flux is remarkably bigger than flood tide flux. As a result, higher amount of sediment is carried out from the radial sand ridges area through the channel system, and thus the stable channel systems are possible to be maintained.

- **Wind and Wave**

The wind direction and magnitude is shown in Figure 6.20. The data is mainly from NOAA NCEP/NCAR Reanalysis Dataset (http://www.esrl.noaa.gov/psd/data/gridded/data.ncep.reanalysis.html) during 1948-2008, in addition to measured data in Lvsi station ($N32°08'$,$E121°37'$) during 1960-2001. The wave distribution at this area is dominated by wind-generated waves with only a minor contribution of swell (Ren, 1986). In the offshore area, the mean significant wave height is 1m from the east-southeast and southeast, with a corresponding mean wave period of 4 seconds, while during the winter storm seasons the maximum significant wave height can be 5m from north-northwest and north, with a corresponding mean wave period of 8.5 seconds (Zhu, 2003). Due to sheltering of the sand ridges, wave breaking, and energy dissipation by bottom friction over the complex bathymetry, the wave conditions inside the channels are generally moderate, except during the storm or typhoon period. Storm waves can have significant effects: the sand ridges are flattened by storm waves and recovered during the normal condition, either by tidal current or by normal wind wave motion. Storm waves dont form the ridges, but ruined them. Within 30 years from 1951 to 1981, the 18 times over 34 times of typhoon passed by this area coincided with the spring tide, which caused the abnormally (anomaly) high water level. The wave setup reached 3.81m in Xiao Yang River (where the Xiao Yangkou Harbor located at its estuary, see Figure 6.19) during the typhoon No.14 in August of 1981(Chen, 2008). The storms are assumed to play an important role of reforming the sand ridges.

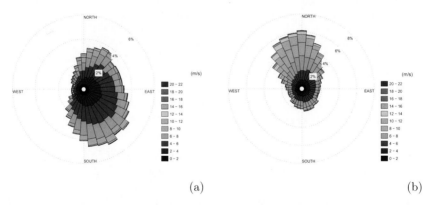

(a)   (b)

Figure 6.20: Wind rose at Lvsi station during 1948-2008, (a) Summer period (May,1 to Oct. 30), (b) Winter period (Nov.1 to Apr. 30)

Sediment characteristics

- **Source of sediment and sediment budget**
  Geological evidence shows that the radial sand ridges are not geological relics. Content analysis shows that the sediment supply is mainly from 2 sources. One source is the fluvial sediments from three adjacent river systems, Yangtze River system, Yellow River system, and Huai River system from the late Holocene. The

other source is the marine sediments by underwater erosion at shoreface/outer shelf and redistribution within the sand ridges and channels (Zhang-RS & Chen, 1992).

Nowadays the part of the fluvial sediment is from Yangtze River, 35 M tons annually transported northwards to the radial sand ridge area by longshore current, and another part is from the small rivers along the coast, 5 M tons per year. Continuous erosion of the huge sediment body of abandoned Yellow river delta transported southwards to the sand ridge area contributes 109 M tons, while the erosion of the Yangtze River delta transported northwards contributes 202 M tons (Wang-Y & Zhu, 1998), and the sediment output budget by the ebb current flow through the channels are estimated as 160 M tons (Zhu, 2003). So the total sediment input budget is around 200 M tons for the whole radial sand ridge area.

Interestingly, the total volume of sand accumulation on the intertidal flat at the landside of radial sand ridges from Sheyang River estuary to Lvsi harbor is estimated 770 M tons, and most of the deposited materials are from adjacent bed and sand ridges (Gao & Zhu, 1988), which indicated that the redistribution of the eroded sediment on the intertidal flat and sand ridges within the area is significant.

- **Bed material**
The bed material sediment distribution was described by Liu and Xia (1983) and Yang (1985). More than 400 box core samples showed that, generally the sediment in the channels is slightly smaller than 100 microns and that on the ridges is a little bit bigger than 100 microns. The d_{50} of the bed material on the West Sun Sand is 0.11mm Lu (2005). Mud content shows strong spatial variation. The profile analysis of Pb-120 for two 70 cm core samples drilled outside of Jianggang in 1992 showed there existed various mud layers in between the sand sedimentation layers Zhang-RS and Chen (1992).

- **Suspended sediment**
Suspended sediment concentration increase from the outer boundary to the landside of the radial sand ridge area, as high as 0.5 kg/m^3 at Lvsi harbor (see Figure 6.19). Seasonal variations are significant. Average concentration at surface layer during winter in the sand ridge area can be as high as 0.3 kg/m^3, while 0.1 kg/m^3 during summer. It is found that the distribution of suspended sediment concentration have strong correlation with the bed material distribution Zhu (2003). The area with coarser bed material has higher suspended sediment concentration, which implies high ambient hydrodynamics. The d_{50} of suspended sediment is 0.02 mm sampled in the West Sun Sand area (Lu, 2005).

Model setup

In addition to field monitoring and in-site survey, processes based modeling has proven advantages to help understanding nature phenomena at large scales. Two approaches have been applied to study shoreface connected sand ridges: i) full process-based modeling, including non-linear processes (Zhang & Zhang, 1996; Zhang et al., 1998; Yan, Zhu, & Xue, 1999; Yan, Song, et al., 1999; Zhu et al., 1995; Zhu, 2003; Chen, 2008), and ii) idealized process-based modeling, such as stability analysis (Trowbridge, 1995; Falqués et al., 1999; Calvete, Falqués, et al., 2001; Calvete, Walgreen, et al., 2001; Calvete et al., 2002; Calvete & De Swart, 2003; Walgreen et al., 2002; De Swart & Calvete, 2003; De Swart et al., 2008).

In this study, a full process-based numerical model is setup using Delft3D, with eight principal astronautical tidal components at the boundaries and wave forces. The bathymetry measured in 2006 is used to calibrate the modal and to verify the hydrodynamic boundaries and input parameter settings. The model results of tidal current and water level at several positions are compared with measurement data and Admiralty Tide table, shown in Section 4.1.

With the calibrated hydrodynamic boundaries and parameters, a morphodynamic model is setup using bathymetry interpolated from historic coastline and bathymetry of 1855 AD. The bathymetry in the deep sea is interpolated from national coastal survey maps in 1979 and 2006 (Chen, 2008; Lu, 2005) and depth data from Yellow Sea Model provided by Deltares (Bijlsma et al., 1998). Furthermore, for the present bathymetry dataset, the bed level and coastlines in the shoreface connected radial sand ridges area and near-field are updated using 1992, 1994, 2003, and 2005 field measurement with the largest map scale of 1:1000. And for historical bathymetry set, the coastal line and bathymetry are schematized based on the contour lines of a historical sketch map made in 1826 AD, supplemented by a map of 1724 AD and a nautical map of 1904 AD by British Navy. The historic maps refer to Chen (2008)(p.105) and Zhang-RS and Chen (1992)(p.14) for details. The model results are analyzed and compared with measured bathymetry in 2006.

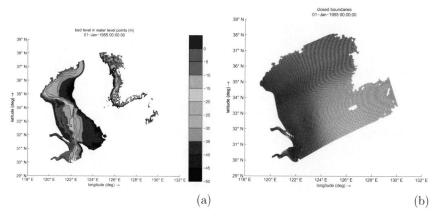

Figure 6.21: Model domain, historical bathymetry. (a) Bathymetry of 1855 AD; (b) Model domain and grid.

Delft3D is a verified numerical modeling tool which fully integrates the effects of waves, currents and sediment transport on morphological development in coastal, river and estuarine areas. The model has been validated in both depth averaged mode (Elias et al., 2000) and three dimension mode (Lesser et al., 2004), including simulation of flow current, sediment transport and morphology change.

Depending on the resolution of available observations, bathymetry data is triangularly interpolated or grid-cell averaged to the curvilinear grid. The well structured, orthogonal curvilinear grid has 40500 points, with maximum resolution 1 km 1.5 km at the center location of radial sand ridge area, Jianggang area (Figure 6.21b). The north boundary is set at the strait between Shandong Peninsula and Korean Peninsula, while the east boundary locates in the middle of Korea Strait, and the south boundary connected the south bank of Qiantang Estuary and Island Kyushu. All the three open-sea boundaries are located "far away" outside the influence sphere of shoreface connected radial sand ridge areas and prescribed as water level boundaries. The water level time series are extracted from the verified Yellow Sea Model (Bijlsma et al., 1998), which is driven by eight principal astronomic tidal components, i.e. Q1, O1, P1, K1, N2, M2, S2 and K2 at the boundaries.

The bed roughness is prescribed as a manning coefficient which is spatial varying as a step function of water depth. The area shallower than 30 m has a value of 0.015, while the deeper area has larger value as 0.02, and the area closed to the west bank of Korean Peninsula with numerous small islands has the value of 0.03 to 0.04.

The secondary flow option is activated to take into account spiral flow due to stream-line curvature caused by the complex bathymetry. The computations start from a restart file generated by pre-run with constant present bathymetry, 720 minutes spin-up prior to the actual morphological computations are used to dissipate the errors induced by the discrepancy between boundary conditions and initial state.

The model is additionally forced by a gridded representative wind time series from 1948-2009 (6 hours interval, 4 times daily) from NOAA NCEP/NCAR Reanalysis Dataset (http://www.esrl.noaa.gov/psd/data/gridded/data.ncep.reanalysis.html). Through statistic analysis on the wind time series, the wind occurrence in 8 representative directions with 3 categories (i.e. $< 8m/s$, between $8m/s$ and $16m/s$, $> 16m/s$) is achieved. The analyzed wind data is used as input for a coarse grid wave model, which is nested with a finer grid wave model. The finer grid wave grid model is aimed to provide wave data to a flow model. Default parameters settings and values are used for the two nested wave models.

Parallel online approach is applied on a Linux cluster (Roelvink, 2006). Different wind conditions at given frequency are input into different flow models. The flow models, coupled with wave models, provide morphological changes to a merging process, which returns a weighted average bottom change to all flow computation processes. The flow models continue the simulation with the updated bottom. Before merging, the bottom changes for each condition can have different morphological factors depending on the statistic wind occurrence.

Table 6.6: Parameter lists and morphodynamic model settings

Module	Parameter	Value	Description
Flow	Δt	10mins	Computational time step
	v_H	$0.1 m^2/s$	Horizontal eddy viscosity
	D_H	$10 m^2/s$	Horizontal eddy diffusivity
Wind	Δt	360mins	6 hours inteval data from NOAA/NCEP
Sediment	D_{50}	0.1mm	Single fraction, non-cohesive bed material
	ρ_s	$2650 kg/m^3$	Sediment density
	d	20m	Initial erodable sediment depth at bed
	$mudcnt$	0.05(-)	Percentage of mud in bed material, discussed later
Morphology	$morfac$	120(-)	Morphological factor

6.4.3 Results

Verification of the flow model

The bathymetry and coastal lines at present time are applied for the verification runs (Figure 6.21a). The river discharge, wind stress and ocean circulation are ignored. The harmonic constants, positions of amphidromic points, tidal current ellipses are used to verify the tidal wave motion at large scale (Figure 6.22). Results from the present model are verified against the results from the validated Yellow Sea Model (Bijlsma et al., 1998). The harmonic constants of K1, S1, M2 and S2 are shown in co-tidal charts (Figure 6.23). The positions of amphidromic points for K1 and M2 tide components are verified against the previous studies from Chen (2008), Yanagi et al. (1997) and Larsen et al. (1985) (Table 6.7). The harmonic constants of K1 and M2 tidal are verified against the measurement datasets in 4 stations in this area (Table 6.8). The tidal current magnitudes and directions are verified against field measurement during various periods (Figure 6.24).

- **Tidal wave motion**
  Tidal wave motions at large scale are demonstrated by co-tidal charts of K1, S1,

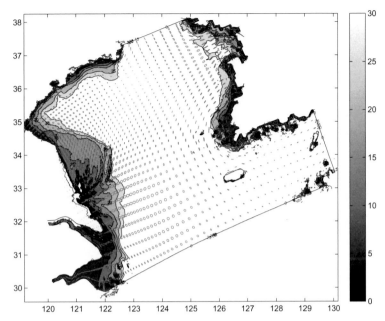

Figure 6.22: M2 tidal ellipse with bathymetry and coastline at present situation
(bathymetry and coastline are based on data of 2006).

M2 and S2 components, which are compared with co-tidal charts generated after the Yellow Sea Model (Bijlsma et al., 1998). The results show quite a good correspondence in both amplitude and phases from the two models (Figure 6.23).

Table 6.7: Position of amphidromic points for M2 and K1

Author	M2	K1	Data source
Choi (Larsen, 1985; Ogura, 1933)	$N35^o$, $E121.5^o$	$N34^o$, $E123^o$	from model
Yanagi (1997)	$N34.5^o$, $E122^o$	$N34.2^o$, $E122^o$	from observation
This paper	$N34.79^o$, $E121.43^o$	$N34^o$, $E123.2^o$	from model

Two M2 amphidromic points showed in this model area are located at $N34^o79'$,$E121^o43'$ and $N37^o6'$,$E123^o35'$ (Figure 6.23c,d), which are comparable with the harmonic analysis results from TOPEX/POSEIDON altimetric data (Yanagi et al., 1997) and Chois model (Choi, 1990) simulation results from Larsen et al. (1985). Table 6.7 listed position of the M2 amphidromic point close to the radial sand ridge area from our model, Chois model (referred by Larsen et al. (1985)) and observed data (Yanagi et al., 1997).

The harmonic constants of K1 and M2 tidal are verified against the measurement

157

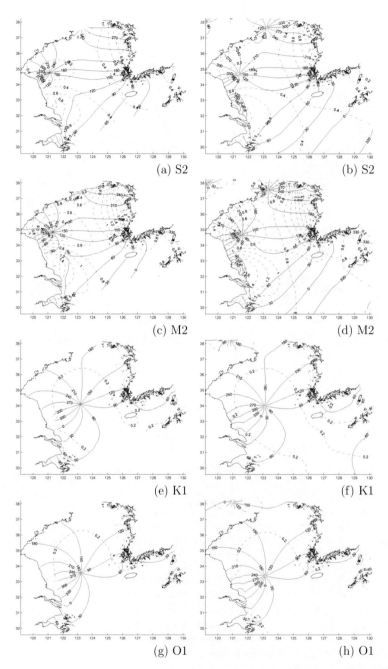

(a) S2   (b) S2

(c) M2   (d) M2

(e) K1   (f) K1

(g) O1   (h) O1

Figure 6.23: S2, M2, K1 and O1 co-amplitude and co-phase charts, left panels are results from present model, and right panels are results from the validated YSM (Bijlsma et al., 1998).

datasets from Admiralty Charts in 4 stations (see Figure 6.19). Table 4 listed the detailed names and coordination. The RMSE (root mean square error) of amplitude is 4.24 cm, and phase lag is 12.12^o for M2 component, except the Yangkou Harbor, and amplitude of 4.74 cm, phase lag of 7.24^o for K1 component. In general, the model results are favorably with the measurement in such a large scale, even though the phase lag errors are somewhat high in the relative shallow area (for example, at Yangkou Harbor), due to the complex bathymetry. A model with higher resolution of bathymetry might help.

- **Tidal current**
  Inside the radial sand ridge area, there are some measurement data at several points for certain periods. The time series of, between 5:00 AM on 24th May, 9:00 AM on 25th May of 2005, velocity magnitude and current direction at two points are chosen for verification.

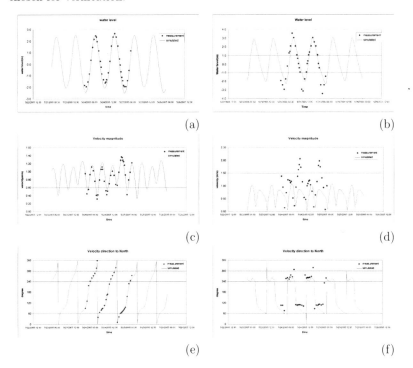

Figure 6.24: The measured and model simulated water level (a,b), velocity magnitude (c,d), velocity direction to North (e,f). Figures at left panel (a, c, e) are measured and simulated results at Point ($N32^o49.052'$, $E121^o58.268'$), while those at right panel (b, d, f) are at Point ($N32^o36.187'$, $E121^o19.316'$).

It is shown that the simulated result succeed to reproduce the water level, tide current vector at one point ($N32^o49.052'$, $E121^o58.268'$) (Figure 6.24a,c,e). However,

Table 6.8: Measured and simulated harmonic constants of M2[1]

Position	Coordination	M2					
		Measured		Computed		Error	
		Amplitude	Phase	Amplitude	Phase	Amplitude	Phase
Qingdao	$N36.08^o$, $E120.32^o$	125	136	120	120	5	16
Yangkou	$N32.6^o$, $E120.93^o$	254	9	194	13	60	-4
Lvsi	$N32.13^o$, $E121.58^o$	175	356	170	7	5	-11
Buoy at Mid Yellow Sea	$N35.6^o$, $E123.7^o$	83	63	81	55	2	8

[1] amplitude in centimeters, and phase in degrees

Table 6.9: Measured and simulated harmonic constants of K1[2]

Position	Coordination	K1					
		Measured		Computed		Error	
		Amplitude	Phase	Amplitude	Phase	Amplitude	Phase
Qingdao	$N36.08^o$, $E120.32^o$	27	356	31	350	-4	6
Yangkou	$N32.6^o$, $E120.93^o$	21	123	14	122	7	1
Lvsi	$N32.13^o$, $E121.58^o$	21	152	17	152	4	13
Buoy at Mid Yellow Sea	$N35.6^o$, $E123.7^o$	14	302	17	300	-3	2

[2] amplitude in centimeters, and phase in degrees

the flow at the other point ($N32^o36.187$', $E121^o19.316$') (Figure 6.24b,d,f), where the water depth is quite shallow, is not perfectly reproduced. The low water level is underestimated, while the peak velocity at ebb time is much lower, even though the phase is correctly reproduced. It occurred typically when the tidal prism volume for intertidal flat in the model is lower than the real case. It is found that the grids are still too coarse to represent the detailed flow and water volume in the cell. In that same grid, the real bathymetry varies from 3 m to 10 m, which is averaged for computation in the model. To reproduce the detailed flow structure, a model with finer grid might help.

From the analysis of co-tidal charts, positions of amphidromic points and harmonic constants of M2, K1 tidal component, we conclude that the tidal wave pattern is reproduced by the model. The verifications of tidal current time series also support that the imposed boundaries and parameters settings are reasonably good even though the finer grid model might increase its performance. The model is well validated for studying the generation and evolution of shoreface connected radial sand ridges.

In summary, from the analysis of co-tidal charts, positions of amphidromic points and harmonic constants of M2, K1 tidal component, we conclude that the tidal wave motion is well reproduced by the model. The verifications of tidal current time series also support that the imposed boundaries and parameters settings are reasonably good even though the finer grid model might increase its performance. The model is well validated for studying the generation and evolution of shoreface connected radial sand ridges.

Morphological modeling

Compared the historic bathymetry (Figure 6.25a) and coast line to the present situation (Figure 6.19 and Figure 6.26b) , two main features can be observed. One is that the abandoned Yellow River delta has been totally eroded away, accompanied with the retreat of coastline at the abandoned Yellow river estuary. The other is that the developments of radial sand ridges, with the coastline extension seawards at this area. Most morphological changes take place within -30 m depth contour. A group of shoals could be observed already in 1855 AD, even though the channels between them are not clearly defined. One big shoal connects with the coastline at the present root position (at Jianggang). In 1855 AD, the sand ridges are at their initial stage of evolution towards the present radial shape.

Figure 6.26a shows the models simulation after 120 years morphological evolution, starting from the bathymetry at 1855 AD. Compared with present bathymetry (Figure 6.26b), it is found again that tide flow is not the only driving force to form the morphological features. With only tide force, the -15m to -20m deep channels between the sand ridges are formed, but the sand ridges are connected to with shore face with an angle and parallel to each other. With wave setting, the overall planar radial shape of the sand ridges is reproduced correctly after some 20 to 40 years in the model (Figure 6.25b, c). In this area, the dominant north wind waves stir up sediment in the underwater delta and wind-driven currents bring suspended sediment southward to the radial sand ridges area. After 40 years evolution in the model, the root position of the sand ridges is correctly captured by the model. Radial shapes of sand ridges with deep channel between them are developed. The directions of the sand ridges look similar to the current situation. However, after 120 years simulation in the model, the under water river delta is removed and reshaped by the hydrodynamics and a longshore sand bar is formed from the model results. It is also found that the outer side areas of the sand ridges are under strong erosion. The cross shore profile at north, middle, and south part is shown as the coast lines along the north and middle part are retreating and south part is under sedimentation. The sand ridge volumes keep decreasing from 1855 then on. The overall morphology adapt to the tidal flow pattern at large scale.

These very preliminary modeling results show that, given historic coastline and bathymetry in 1855, with the given sediment supply and tidal boundary conditions, the main morphological features shown nowadays might be reproduced qualitatively. It implies that the tide and waves are the mechanisms responsible for the generation and evolution of this sand ridge system. Their spatial shape nowadays is the result of a combination of initial bathymetry and the tidal and wave system, but not only one of them.

Starting from the historical bathymetry of 1855 AD (Figure 6.21c and d), morphodynamic modeling for 120 years is carried out. The preliminary results of every 20 yrs morphology demonstrate detailed morphological responses to hydrodynamics (tide and wave) (Figure 6.25). The simulated spatial patterns show a trend that is in reasonable agreement with measurements in 2006. The results confirm the hypothesis of previous studies that the unique radial shape of the sand ridges is mainly caused by the radial tidal wave pattern in this area (Zhang and Zhang, 1996). Moreover, because the suspended sediment transport plays a more important role than bedload transport in this area, wave effects are indispensable to the suspended sediment transport processes and the subsequent morphological evolution.

To replicate the morphological changes in 150 years within a feasible simulation time, forcing conditions are reduced through the use of an input reduction method. Two morphological tides are applied at boundaries. Sixty years measured wind conditions are reduced to 8 categories, which are input into a nested SWAN model. The computed wave characteristics are coupled with flow and morphology models. Parallel online approach (Roelvink, 2006) is applied to evaluate the morphology changes at corresponding wind wave probabilities.

6.4.4 Discussions

Morphodynamic modeling

The hydrodynamic modeling results presented here compare well to other modeling studies. For the shoreface connected radial shape sand ridges at southern Yellow Sea, Zhang and Zhang(1996, 1998), and also Zhu et al (1995) proposed that tide current is the dominant force. Although the radial tidal wave pattern at the radial sand ridge area is demonstrated by their numerical model, their approach has assumed that flow is running on an existing radial topography. They also hypothesized that, with minor effects by the local bathymetry, the strong tidal currents could reform the huge sand body into this remarkable radial shape. These conclusions had been adapted by Yan et al (1999a, b) using similar approach. Moreover, Yan et al (1999a, b) addressed that the differences of tidal wave energy rate and current strength determined the different planar shapes of ridges and troughs in the north and the south flank. Chen (2008) assumed that the outer sea boundary of the radial sand ridges is under serious erosion and sea level rise in a constant rate. These two factors result in a larger tidal volume in this area, and consequently the tidal range gets higher, and thus the tidal wave propagation pattern gets slightly different to the present pattern. However, in all these studies, the bathymetry and coast lines do not change during the modeling period. No sediment transport and morphological feedback to hydrodynamics is considered, which is complemented by this paper. Additionally, wave effects on sediment transport and morphological evolution are

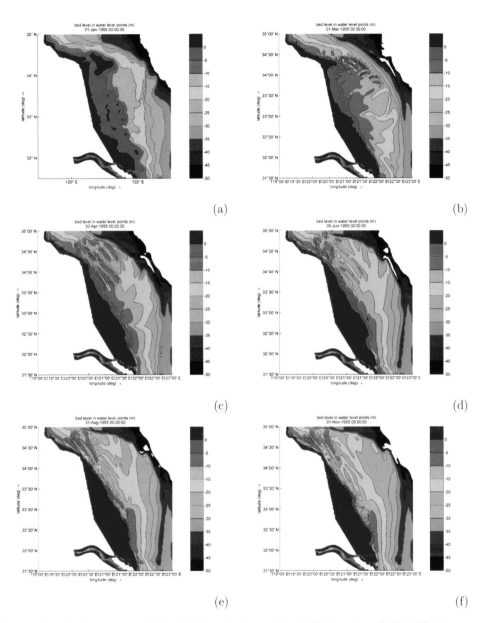

Figure 6.25: Bathymetry evolution with time in the model. (a) Bathymetry of 1855 AD, the inital bathymetry for model; (b) Morphology after 20 years; (c) After 40 years; (d) After 60 years; (e) After 80 years; (f) After 100 years;

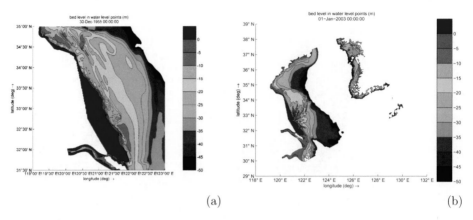

(a)   (b)

Figure 6.26: (a) Simulated bed level after 120 years; (b) Measured bed level at 2006

excluded in the previous studies. In this paper, the non-linear responses among flow, wave, sediment transport and morphological processes are taken into accounted through the online coupling technique. In each time step, the bathymetry is updated and used in the subsequent time step of the hydrodynamic computation. Considering the different scales of hydrodynamic and morphological processes, the morphology updating is accelerated by morphological factor (Roelvink, 2006).

The validation of the tidal current at several observation points showed some discrepancies between the model simulation and measurements, especially in the shallow water areas. The reason is that computation grid is still to coarse to have accurate the tidal prism volume in the shallow areas. And also energy dissipation by bottom friction is underestimated. A local model with higher resolution might help to increase the validation accuracy of the detailed flow velocity and sediment concentration. The radial sand ridges constitute a huge scale system of hundreds of kilometers. It could be divided into several subsystems which include subsets of morphology features, such as one ridge with adjacent channels. To identify the boundaries of the sub-morphological features, a detailed model might be necessary. For subsystems with smaller spatial scale, compared to the whole huge system, the morphological changes in relatively short term are also recognizable and might be easier to understand. The present study could provide boundary conditions for further studies on the subsystems. It is also suggested that idealized process-based modeling approach, such as stability analysis, could be applied, as a compliment for the long term, large scale morphological features (Idier and Astruc, 2003). Considering the shoreface connected radial shape sand ridges as rhythmic features, stability analysis has been applied by many researchers. Trowbridge (1995) predicted the ridges growth in the inner shelf with correct orientation and shape by using shallow water equations with sediment mass continuity equation and sediment transport flux as a linear function of velocity. This method was extended by Falques et al (1998) with Coriolis forces and diffusive sediment transport flux. Calvete et al (2001) extended the method by including suspended sediment transport, bed load transport and wave stirring effects. Walgreen et al (2002) included tidal asymmetry effects and enhanced suspended sediment transport during storms. All the studies demonstrated that the initial growth of the ridges can be explained as the inherent free instabilities of the coupled fluid bottom system. Calvete et al (2002) and Calvete and De Swart (2003) extended the model to the non-linear regime to reveal the final shapes of the bed forms, spatial patterns, migration speeds and equilibrium time scales. More physical processes were taken account for, such as, alongshore pressure gradient, non-equilibrium suspended sediment transport. The non-linear stability analysis model was applied to study the human interventions, like sand extracting, nourishment and construction of navigation channels (De Swart and Calvete, 2003). Swart et al (2008) discussed the impact of grain sorting and applied the model to La Barrosa beach in Spain and Long Island shelf in US. All the applications applying stability analysis are limited in the storm-dominated inner shelves with micro- or meso-tidal range, which is obviously not the case of the shoreface connected radial sand ridges at southern Yellow Sea. Further investigations are required to elaborate on the mechanisms behind the evolution of this shoreface connected radial shape sand ridges.

Radial tidal wave pattern

Radial tidal wave pattern is formed by combination of the two tidal waves with mild slope continental shelf.

A schematized model is set up to check this hypothesis. A rectangular model domain covers an area of 500 × 900 kilometers by 12.5 × 12.5 kilometers grid cells to represent the whole East China Sea (Figure 6.27a). A M2 tide wave propagates from the only open boundary at the south of the domain. If the shelf bathymetry is designed as a mild linear slope at the cross section direction from (linear slope from 0m to 70m and then another linear slope from 70m to 300m), the M2 tidal ellipse shows that even with this highly schematized model settings, the radial shape of tidal wave pattern still exists (Figure 6.27b). However, if the shelf bathymetry is set as a horizontal flat bed with uniform depth of 30 meters (Figure 6.28a), the radial shape of tidal wave does not exist (Figure 6.28b).

Relation of radial tide wave and radial sand ridge

There is no clear evidence showing the radial sand ridge is formed by the radial tidal wave. There might be other mechanisms, such as wave effects, or disturbations by storm surge, combining with tidal force to form these huge radial sand ridges.

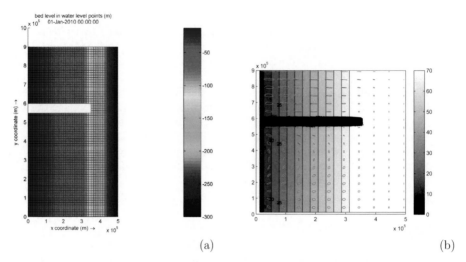

Figure 6.27: Schematized model domain (a) bathymetry (b) M2 tidal ellipse with SLOPED bed

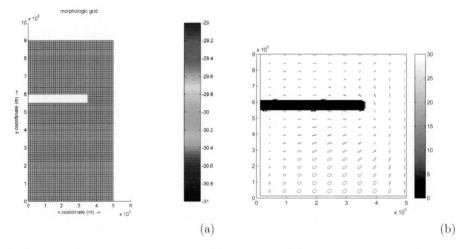

Figure 6.28: Schematized model domain (a) bathymetry (b) M2 tidal ellipse with FLAT bed

6.4.5   Conclusions

This section presents the morphodynamic simulations using a full processed based model for the unique shoreface connected radial sand ridges located at southern Yellow Sea, China, during 1855-2006 period. After calibration and validation of tide wave propagation using the present bathymetry, the model is used to investigate the initial formation from a schematized flat bed firstly. Secondly the model is applied to examine the morphological evolution based on the historic bathymetry from 1855 to 2006.

The validation runs with present bathymetry successfully reproduced the radial tidal wave pattern formed by the progressive tidal wave from Pacific Ocean meeting the tidal wave in anti-clockwise rotation in the radial sand ridge area. The cotidal charts show good correspondence in both amplitude and phases with results from the validated Yellow Sea continental shelf model. Positions of the amphidromic points for M2 and K1 are well comparable with the harmonic analysis results from TOPEX/POSEIDON altimetric data and modeling results from Choi (1990). Besides, the harmonic constants of M2 and K1 tidal are verified against the measurement datasets from Admiralty Charts in 4 stations, which are also favorable in general, except for the Yangkou station at shallow water. The tidal current vectors validated against the field measurements also show good agreement. The validation results show that the boundaries and parameters setting are reasonably good. The model is well validated for studying the generation and long-term evolution of shoreface connected radial shape sand ridges.

The runs with historic bathymetry at 1855 AD show the maintenance of the radial sand ridge with given historic bathymetry and hydrodynamic conditions during the past 120 years. The abandoned Yellow River underwater delta has been totally eroded. The radial sand ridges with deep channels between them are getting distinctive eventually. A longshore sand bar is formed instead.

All model runs reproduce that the outer side shelves of the sand ridges are under erosion. The overall morphology adapt to the tidal flow pattern at large scale gradually. The model runs cannot confirm the hypothesis of previous studies (Zhang & Zhang, 1996; Yan, Zhu, & Xue, 1999; Yan, Song, et al., 1999; Zhu, 2003) that the unique radial shape of the sand ridges is mainly caused by the radial tidal wave pattern in this area. Moreover, due to that the suspended sediment transport plays a more important role than bedload transport in this area, wave effects is added to the subsequent morphological evolution. The initial bathymetry might be important for the morphological evolution in this area as well.

The study provides a large scale framework for morphodynamic evolution of the shoreface connected radial sand ridges located at southern Yellow Sea, China. For morphological processes in smaller scale, such as helical flow as secondary current, graded sediment transport etc., models with finer grid may help. Furthermore, for a complete understanding of the hydrodynamic-morphology interaction, linearized analysis of the stability of the topographical perturbation would be also a valuable compliment in the further study.

This study area is both ecologically and economically important. Although the intention was to test the full bio-geomorphological system on this study area, time didn't allow for it. However, rather than completely excluding this study area from this thesis, we include this first step since our simulations (carried out using standard Delft3D online MOR) shed some new light on the genesis of the sand ridge pattern as it exists today, which hopefully provides a good start for further study on both morphology and ecology in this area.

6.5   Conclusions

In this chapter, the generic geomorphological model is applied in two cases at different temporal and spatial scales. Section 6.2 shows the morphological changes with vegetation effects. The channel network development in tidal salt marsh is focused on. Even with very preliminary settings and minor calibration efforts, the tidal channel pattern within the salt marsh is captured by the model reasonably. The vegetation effects on the channel pattern development are distinguished. Section 6.3 shows the morphology and vegetation development in a salt marsh restoration project at Nisqually river, US. Due to the limitation of field flow and sediment data from the Nisqually River, the model is not validated quantitatively. Nevertheless, the vegetation population dynamic processes are implemented in the model, and spatial patterns of vegetation distribution are reproduced qualitatively. Interactions of physical processes and ecological processes are represented reasonably well. The effects of vegetation processes on the morphological development are significant. With vegetation effect, sedimentation in the river channels decreased and sedimentation increased on the tidal flat area where the vegetation grow and spread. Section 6.4 describes a unique large morphological feature, shoreface connected radial shape sand ridges located at southern Yellow Sea, China. Tidal wave propagation and tide current are validated against measurement and literatures. The generation of the large shoreface connected radial shape sand ridges are regarded as a result of the tide wave system and considerable sediment source in this area. Long-term evolution trends are proposed, which needs to be further validated against field measurements in future.

Chapter 7

Conclusions and Recommendations

7.1   Introduction

The objective of this study is to develop a generic geomorphological model to increase our ability of analysis and prediction of the coastal system based on the increasing understanding of coastal processes and the interactions between them at different temporal and spatial scales.

In this book, the generic geomorphological model is described in chapter 3 and validated in Chapter 4. The adaptive numerical algorithm and the open model structure make this morphological model to be a generic model. The generic geomorphological model is extended to be a bio-geomorphological model by integrating the ecological processes in Chapter 5. The bio-geomorphodynamic processes and their interactions are also verified in Chapter 5. Finally, this generic geomorphological model is applied in three cases at different temporal and spatial scales in Chapter 6. The first one demonstrates morphological changes with vegetation effects in a tidal basin with developing salt marshes, which is in relative small spatial scale (a few kilometers) and short temporal scale (decades). The channel network development in tidal salt marsh is focused on. The second one is a real-world case study on mid term morphological evolution with vegetation effects in a salt marsh restoration project in US. Even though the ecological processes are highly schematized in the model and very limited calibration efforts have been put on, the extended generic geomorphological model still indicates the significance of ecological effects on morphological processes. The applications also demonstrate the potential ability of the generic geomorphological model to be a generic platform for multi-discipline study. The third one studies a unique large morphological feature, shoreface connected radial shape sand ridges located at southern Yellow Sea, China. The long-term evolution trend is focused on.

As a final result, it is clear that the objectives have been met, as a modeling approach for validation cases and applications. Our ability of analysis and prediction of the coastal system at different scales are improved with this flexible modelling tool. To achieve the final objective, a few research questions are to be answered as the following section.

7.2 Development and validations of a generic geomorphological model

How can this model be made generic from a mathematics point of view?

- *How could the generic model be coupled with various hydrodynamic models, no matter if they are based on structured or unstructured grids and irrespective of the numerical integration methods?*

The generic geomorphological model is used to couple with various FLOW models, both for models based on structured grid and unstructured grid. This is assured by its open structure and adaptive numerical schemes.

The model development uses the open processes library, an open frame of Delft3D-WAQ. Open processes library collects hundreds of substances, corresponding processes and parameter sets. Sediment fractions and related morphodynamic processes are added in the library (refer to Chapter 3). Several vegetation species and corresponding ecological processes are integrated in to the library (refer to Chapter 5). For sediment transport, the transport solver in Delft3D-WAQ applies finite volume method, which integrates all sediment transport fluxes passing the exchanges. For the Arakawa C-grid type model, a generic velocity integration algorithm is developed for the velocity vectors in grid center (refer to Section 3.5). Mass conservation is assured in the velocity integration procedure.

For morphological update, a set of numerical algorithms are discussed. And the TVD filter is suggested.

In addition, the extension of the bed state description module (refer to Section 3.2) and sediment (mud) transport (refer to Section 3.3) broaden the applicability of the generic geomorphological model.

- *How can the generic model represent and couple the dominant coastal sediment transport processes and morphodynamic processes, and possibly, ecological processes?*

The involved processes in this generic geomorphological model include sediment transport processes, geomorphological bed level updating, and ecological processes. Sediment transport process and morphological update, ecological process are implemented for both structured grid and unstructured grid (refer to Chapter 3). The scale coupling is implemented by parameterization or by incorporation of a few crucial aspects of dynamics from processes at the next level of (larger or smaller) scale (refer to Section 2.2.1 and Section 5.2.2).

The morphological factors are used to bridge the gap of morphological processes and others. Discussions on the application of morphological factors are discussed when wave processes are involved and ecological processes are involved (refer to Section 5.3.5).

Within this study, the discussion on validation of the generic geomorphological model is limited since the unstructured flow model is not ready. However functionally, the results of the generic geomorphological model have been validated against the results from existing analytical solutions, flume experiments and other verified numerical simulation results, such as, from Delft3D-FLOW online version and Telemac (refer to Chapter 4). This new model gives out equally good results as other processed-based models with similar parameters settings. Furthermore, this new model offers more, i.e., improved processes, capable for unstructured grid, which make that this model has more potential towards a generic platform for multi-discipline study.

Is this model is generic from an application point of view?

- *Could this generic morphological model reproduce typical morphological phenomena at large scales, such as morphological evolution of sand ridge system, and at small scales, for instance, trench migration?*

The generic morphological model is planned to be applied to study the morphology evolution of shoreface connected radial shape sand ridges located at southern Yellow Sea, China. Nevertheless, the power of the generic geomorphological model lies in its flexibility for unstructured grid and extendibity towards biomorphological model. These features are not really highlighted at present stage of this study. We decided to carry out the morphodynamic analysis with the fast and robust present Delft3D online MOR model, as a necessary first step towards fully coupled bio-geomorphological simulations. Results from a first experiment show that even though morphological processes are rather complex, the generic geomorphological model still provides rather similar morphological patterns compared to the results from the Delft3D online MOR (refer to Figure 6.18). Preliminary results from the Delft3D online MOR model shows that the tidal wave motion, current pattern from the model results well reproduced compared to measurements and previous study. Harmonic tide component and tidal current are also well reproduced by the model (refer to Section 6.4.3).

This generic morphological model also reproduces typical morphological phenomena at small scales. One example is trench migration, which is simulated by this model. The model result is validated again flume experiments and model results from Delft3D online MOR model. The trench migration speed and phase are well reproduced by this model (refer to Section 4.3.1).

Another example is tidal channel development in salt marsh. The model shows that, with vegetation effects, the tidal channel pattern in salt marsh is much different to the pattern without vegetation effects.

- *Could this generic morphological model reproduce small scale ecological effects on morphological changes in highly dynamic environment, such as salt marshes?*

The generic morphological model is extended to be a bio-geomorphological model by integrating the ecological processes, i.e., vegetation population dynamics. The vegetation population dynamic processes include growth, mortality, lateral expansion (spatial spreading), competition/interaction among species, seed dispersion (refer to Section 5.2). The ecological processes are validated against the observation data in Lake Veluwe, NL (refer to Section 5.2). Even though the parameter sets and relevant processes are still instinctive, the micro scale morphological features in a schematized salt marsh are well predicted.

7.3 Recommendations

Development and validation of the generic geomorphological model

The development of the generic geomorphological model inherits enormous genetic advantages from the Delft3D online MOR version and extends a number of physical-based processes, for instance, the bed form predictor, the unified form of source and sink term to count the mud and sand transport, etc. However, there is a list of processes still missing in this system, such as, transport of graded sediment, the bed roughness prediction etc., which are worthwhile for future study.

The validation of the generic geomorphological model is not thoroughly finished. The coupling with the 1D/2D flow model Sobek is still ongoing. The coupling with flow model using unstructured grid, such as, Telemac, DFlow-FM is also our future plan.

Towards a bio-geomorphological modeling tool

Even though some micro scale morphological features in a schematized salt marsh are reproduced with instinctive parameter sets, it is strongly recommended to integrate more ecology knowledge in the model. Vegetation effects on hydrodynamics have been studied intensively during recent years. However, the interaction of vegetation and sediment transport and morphodynamics needs more attention. Furthermore, vegetation population dynamics in this study are limited to only stiff (cylinder-like) vegetations. In reality, a lot of species of vegetation are quite flexible or half stiff, which needs more investigation. Also the biota effects on the sediment transport, especially on mud entrainment, consolidation etc., are of great interests. Biota population dynamic processes are on a relatively smaller scale and more dynamic compared to vegetation related processes, thus more insight is needed for coupling between the processes. Gradually the model may be developed to be a decent bio-geomorphological modelling tool.

The mechanism of the generation and development of the large scale Shoreface Connected Radial Sand ridges at southern Yellow Sea, China

In this study, the mechanism of the generation and development of the large scale Shoreface Connected Radial Sand ridge at southern Yellow Sea, China are studied using process-based numerical model with schematized settings. But for even longer term of morphological evolution, the stability analysis would be a powerful tool. Even though a number of schematizations are necessary to carry stability analysis for this problem, which might limit the engineering application range, the results of stability analysis are expected to set a boundary for the long term morphological change for this large scale morphological feature.

Appendix A

Process library definition for the morphology model

Calc general fluid parameter

	Water	
Z1		procID: A_aqaParam

Input:
Temp of ambient water body (def: 15°C)
Salinity at zero chloride concentration (0.03g/kg)

Output:
Density of water (kg/m3)
Chloride in water
Kinematic viscosity (m2/s)

Calc general fluid parameter

	Water	
Z2	procID: A_SndParam	

Input:
D10
D35
D50
D90
Dmean
Density of sed
Density of water
G
Kinematic viscosity
Wave period
Size of clay fraction in percentage (Pmud)
Total depth of water column
Depth segment
Velocity at segment center
Velocity direction at segment center

Output:
Critical shear stress for deposition
Critical depth-average velocity for sedB
Critical peak orbital velocity for sedB
Mobility parameter for sedB
Suspended sand size for sedB

Calc sediment deposition velocity

	SedB0i	
H		procID: DepV_S0i

Input:
Density of SedB0i
Density of water (from Z1)
G
Diameter of sedB0i (from A)
Kinematic viscosity (from Z1)

Output:
Settling velocity for SedB0i (m/s)

Calc general Sed parameter

	Sed for all fractions	
Z3		procID: A_FricPara

Input:
Density of sedB01
Density of water
G
D10
D35
D50
D90
Velocity at segment center
Velcoity direction at segment center
Total depth water column
Depth of segment
Calibration coef. for bed roughness
Wind wave height
Wind wave length
Wind wave period
Kinematic viscosity

Output:
Overall Current related bed roughness (m)
Current related bed roughness due to ripples (m)
Current related bed roughness due to Mega ripples (m)
Current related bed roughness due to River dunes (m)
Wave related bed roughness due to ripples (m)
Apparent bed roughness (m)
Grain related Chezy (m0.5/s)
Overall current related Chezy (m0.5/s)
Grain related friction factor (current related) (-)
Overall friction factor (current related) (-)
Apparent friction factor (-)
Grain related friction factor (wave related) (-)
Overall friction factor (wave related) (-)

Calc general friction parameter

	Sed for all fractions	
Z4	procID: A_BFParam	

Input:
Velocity at segment center (m/s)
Velocity direction at segment center to the east (rad)
Calculated height of a wind induced wave (m)
Calculated length of a wind induced wave (m)
Calculated period of a wind induced wave (s)
Total bottom shear stress (N/m2)
Critical shear stress for sedimentation SedB (N/m2)
D_10 (m)
D_35 (m)
D_50 (m)
D_90 (m)
Total depth water column (m)
Depth of segment (m)
Gravitational acceleration (m/s2)
Density of SedB01 (kg/m3)
Fluid density from General fluid param (kg/m3)
Kinematic viscosity from General fluid param (m2/s)

Output:
height of Sndwave/ripple/MegaR/Dune (m)
Wave length of Sndwave/ripple/MegaR/Dune (m)
Bedform Type(1=sndwave,2=ripple,3=mgrppl,4=dune) (-)
Location(1=river,2=estuary,3=coastal) (-)
Flow regime (1=lower,2=high,3=trsit regime) (-)

Figure A.1: Description of the processes for parameters

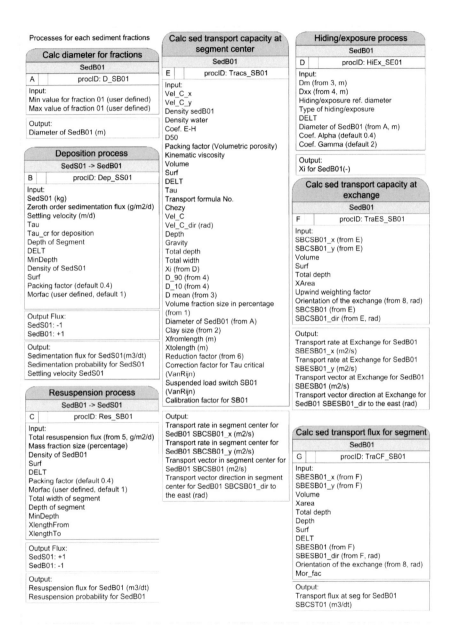

Figure A.2: Description of the processes used by each sediment fraction

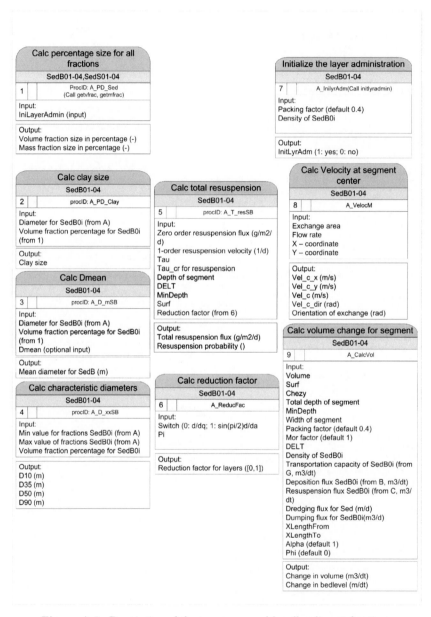

Figure A.3: Description of the processes used by all sediment fractions

References

Aagaard, T., Kroon, A., Andersen, S., Srensen, R. M., Quartel, S., & Vinther, N. (2005). Intertidal beach change during storm conditions; egmond, the netherlands. *Marine Geology, 218*, 65-80.

Antia, E. (1996). Patterns of tidal flow asymmetry on shoreface-connected ridge topography off spiekeroog island, german bight. *Ocean Dynamics, 48(1)*. doi: 10.1007/BF02794055

Atkinson, J., Westerink, J. J., & Hervouet, J. M. (2004). Similarities between the quasi-bubble and the generalized wave continuity equation solutions to the shallow water equations. *INTERNATIONAL JOURNAL FOR NUMERICAL METHODS IN FLUIDS, 45*, 689-714.

Bagnold, R. (1966). An approach to the sediment transport problem from general physics. *Geol. Surv. Prof. Paper, US Geological Survey, DOI, USA, 422-I.*

Bakker, W., Klein Breteler, E., & Roos, A. (1970). The dynamics of a coast with a groyne system. *In: Procedings of the 12th Coastal Engineering CONFERENCE, Sept. 13-18, 1970, WASHINGTON, D C, VOLUME 2; ASCE, NEW YORK, 23 FIG, 12 REF, APPEND*, 1001-1020.

Baptist, M. (2005). Modeling floodplain biogeomorphology. *Phd Dissertation of TU Delft University*, 154.

Battjes, J., & Jansen, J. (1978). Energy loss and set-up due to breaking of random waves. *Proc. 16th Int. Conf. on Coastal Engineering. ASCE, New York*, 570- 587.

Bayram, A., Larson, M., Miller, H. C., & Kraus, N. C. (2001). Cross-shore distribution of longshore sediment transport: comparison between predictive formulas and field measurements. *Coastal Engineering, 44*, 79-99.

Bijlsma, A., Graaff, R. d., & Thabet, R. (1998). Numerical modelling for the changes of tidal characteristics in the nearshore of barrier dams. *Final Engineering report, H3086, WL — Delft Hydraulics.*

Booij, N., Ris, R., & Holthuijsen, L. (1999). A third generation wave model for coastal regions: 1. model description and validation. *J. Geophys. Res., 104*, 7649- 7666.

Borsje, B., De Vries, M., Hulscher, S., & De Boer, G. (2008). Modeling large-scale cohesive sediment trans-port with the inclusion of small-scale biological activity. *Estuarine, Coastal and Shelf Science, 78*, 468-480.

Bouma, T., Van Duren, L., Temmerman, S., Claverie, T., Blanco-Garcia, A., Ysebaert, T., & Herman, P. (2007). Spatial flow and sedimentation patterns within patch of epibenthic structures: combining field, flume and modeling experiments. *Continental Shelf Research, 27*, 1020-1045.

Briand, M., & Kamphuis, J. (1993). Sediment transport in the surf zone: a quasi 3-D numerical model. *Coastal Engineering, 20*, 135-156.

Brøker Hedegaard, I., Roelvink, J., Southgate, H., Pechon, P., Nicholson, J., & Hamm, L. (1992). Intercomparison of coastal profile models. *In: Proceedings, 23rd International Conf. on Coastal Engineering, Venice. Vol. 2, pp. 2108-2121..*

Burg, M., & Tripp, E., D.R.and Rosenburg. (1980). Plant associations and primary productivity of the nisqually salt marsh on southern puget sound. *Northwest Science, 54*, 222-236.

Burningham, H., & French, J. (2006). Morphodynamic behaviour of a mixed sand-gravel ebb-tidal delta: Deben estuary, suffolk, UK. *Marine Geology, 225*, 23-44.

Callaghan, D., Saint-Cast, F., Nielsen, P., & Baldock, T. (2006). Numerical solutions of the sediment conservation law; a review and improved formulation for coastal morphological modelling. *Coastal Engineering, 53*, 557-571.

Calvete, D., & De Swart, H. (2003). A nonlinear model study on the long-term behavior of shoreface-connected sand ridges. *J. Geophys. Res., 108(C5)*, doi:10.1029/2001JC001091.

Calvete, D., De Swart, H., & Falqués, A. (2002). Effect of depth-dependent wave stirring on the final amplitude of the shoreface-connected sand ridges. *Cont.Shelf Res.*, *22*, 2763-2776.

Calvete, D., Falqués, A., De Swart, H., & WALGREEN, M. (2001). Modelling the formation of shoreface-connected sand ridges on storm-dominated inner shelves. *J. Fluid Mech.*, *441*, 169-193.

Calvete, D., Walgreen, M., De Swart, H., & Falqués, A. (2001). A model for sand ridges on the shelf: Effect of tidal and steady currents. *J. Geophys. Res.*, *106(C5)*, 9311-9325.

Cañizares, R., Alfageme, S., & L., I. J. (2003). Modeling of morphological changes at shinnecock inlet, new york, USA. *Proceedings, Coastal Sediments 2003, American Society of Civil Engineers.*

Cayocca, F. (2001). Long-term morphological modeling of a tidal inlet: the arcachon basin, france. *Coastal Engineering*, *42*, 115-142.

Chen, K. (2008). Research of land and ocean interaction process of the coastal zone in jiangsu province since the yellow river flow northwards. *Ph.D. dissertation of Nanjing Hydraulic Research Institute.*

Chen-C, Liu, H., & Beardsley, R. (2003). An unstructured grid, finite volume, three-dimensional, primitive equations ocean model: application to coastal ocean and estuaries. *Journal of Atmospheric and Oceanic Technology*, *20(1)*, 159-186.

Choi, B. (1990). A fine grid three dimensional m2 tidal model of the east china sea. *Modelling Marine System, Alan M. Davis (eds.), CRC press,*, 167-185.

Coeffe, Y., & Pechon, P. (1982). Modelling of sea-bed evolution under waves action. *Proc. 18th Int. Coastal Eng. Conf. Cape Town*, 1149-1160.

Cohen, J. (1960). A coefficient of agreement for nominal scales. *Educ. Psychol. Meas.*, *20*, 37-46.

Computation of wind-wave spectra in coastal waters with swan on unstructured grids. (2010). *Coastal Engineering*, *57(3)*, 267 - 277.

D'Alpaos, A., Lanzoni, S., Marani, M., & Rinaldo, A. (2007). Landscape evolution in tidal embayments: Modeling the interplay of erosion, sedimentation, and vegetation dynamics. *J. Geophys. Res.*, *112*, F01008, doi:10.1029/2006JF000537.

Dalrymple, R., Kirby, J., & Hwang, P. (1984). Wave diffraction due to areas of energy dissipation. *J. Waterw. Port Coast. Ocean Eng.*, *110*, 67-79.

Davies, A., van Rijn, L., Damgaard, J., van de Graaff, J., & Ribberink, J. (2002). Inter-comparison of research and practical sand transport models. *Coastal Engineering*, *46*, 1-23.

Dean, R. (1977). *Equilibrium beach profiles: Us atlantic and gulf coasts* (Technical Report No. Report No: Ocean Engineering Report No 12). Newark, Delaware: University of Delaware.

Deigaard, R., & Fredsøe, J. (2002). The shape of equilibrium coastlines [report].

Deltares. (2010a). *Delft3D-FLOW,User Manual* (3.14.12556 ed.). Rotterdamseweg 185, P.O. Box 177, Delft, The Netherlands.

Deltares. (2010b). *Delft3D-WAQ,User Manual.* Rotterdamseweg 185, P.O. Box 177, Delft, The Netherlands.

Deltares. (2010c). *Delft3D-WAVE,User Manual.* Rotterdamseweg 185, P.O. Box 177, Delft, The Netherlands.

De Swart, H., & Calvete, D. (2003). Non-linear response of shoreface-connected sand ridges to interventions. *Ocean Dynamics*, *53*, 270-277.

De Swart, H., Walgreen, M., Calvete, D., & Vis-Star, N. (2008). Nonlinear modelling of shoreface-connected ridges: Impact of grain sorting and interventions. *Coastal engineering*, *55*, 642-656.

Development of a three-dimensional, regional, coupled wave, current, and sediment-transport model. (2008). *Computers & Geosciences*, *34*(10), 1284 - 1306. (¡ce:title¿Predictive Modeling in Sediment Transport and Stratigraphy¡/ce:title¿)

De Vriend, H. (1987). 2DH mathematical modelling of morphological evolutions in shallow water. *Coastal Engineering*, *11*, 1-27.

de Vriend, H. (1990). Morphological processes in shallow tidal seas. In R. Cheng (Ed.), *Residual currents and long term transport, coastal and estuarine studies* (pp. 276-301,vol. 38). Berlin: Springer.

De Vriend, H. (1991). Mathematical modelling and large-scale coastal behaviour, part i: Physical processes. *Journal of Hydraulic Res.*, *29*(6), 727-740.

De Vriend, H. (2001). Long-term morphological prediction. In S. G. & P. Blondeaux (Eds.), *River, estuarine and coastal morphology*.

de Vriend, H., Capobianco, M., Chesher, T., de Swart, H., Latteux, B., & Stive, M. (1993). Approaches to long-term modelling of coastal morphology: a review. *Coastal Engineering*, *21*, 225-269.

De Vriend, H., & Stive, M. (1987). Quasi-3D modelling of nearshore currents. *Coastal Engineering*, *11*, 565-601.

De Vriend, H., Zyserman, J., Nicholson, J., Roelvink, J., Pchon, P., & Southgate, H. (1993). Medium-term 2dh coastal area modelling. *Coastal Engineering*, *21*, 193-224.

De Vries, M., & Borsje, B. (2007). Species response to climate could affect the fate of fine sediment in the wadden sea. In Dohmen-Janssen & Hulstcher (Eds.), *River, coastal and estuarine morphodynamics* (p. 33-86). Taylor&Francis Group, London, ISBN 978-0-415-45363-9.

Dietrich, J., Zijlema, M., Westerink, J., Holthuijsen, L., Dawson, C., Luettich, R., ... Stone, G. (2011). Modelling hurricane waves and storm surge using integrally-coupled, scalable computations. *Coastal Engineering*, *58*, 45-65.

Dijkstra, J., & Uittenbogaard, R. (2010). Modeling the interaction between flow and highly flexible aquatic vegetation. *Water Resour. Res.*, *46*, *W12547*.

Dingemans, M. (1997). *Water wave propagation over uneven bottoms. part 1 -linear wave propagation*. Advanced Series on Ocean Engineering, 13. World Scientific, 471. pp.

Dissanayake, D., Roelvink, J., & van der Wegen, M. (2009). Modelled channel patterns in a schematized tidal inlet. *Coastal Engineering*, *56*, 1069-1083.

Drønen, N., & Deigaard, R. (2007). Quasi-three-dimensional modelling of the morphology of longshore bars. *Coastal Engineering*, *54*, 197-215.

Dronkers, J. (2005). *Dynamics of coastal systems* (Advanced Series on Ocean Engineering, 25 ed.). NJ (USA): World Scientific: Hack-ensack, ISBN 981-256-349-0.

Einstein, H. (1950). The bed load function for sediment transportation in open channel flow. *Technical bulletin , U.S. Dep. of Agriculture, Washington, D.C., no. 1026*.

Elias, E. (2006). Morphodynamics of texel inlet. *Ph.D. dissertation of TU Delft University*, 262. (ISBN 1-58603-676-9)

Elias, E., & Van Der Spek, A. (2006). Long-term evolution of texel inlet and its ebb-tidal delta (the netherlands). *Marine Geology*, *225*, 5-21.

Elias, E., Walstra, D., Roelvink, J., Stive, M., & Klein, M. (2000). Hydrodynamic validation of delft3d with field measurements at egmond. *27th International Conference on Coastal Engineering, ASCE, Sydney*.

Falqués, A. (2003). On the diffusivity in coastline dynamics. *Geophys. Res. Lett.*, *30(21)*, 2119, doi:10.1029/2003GL017760.

Falqués, A., & Calvete, D. (2005). Large-scale dynamics of sandy coastlines: Diffusivity and instability. *Journal of Geophysical Research*, *110*(C03007), 15.

Falqués, A., Coco, G., & Huntley, D. A. (2000). A mechanism for the generation of wave-driven rythmic patterns in the surf zone. *Journal of Geophysical Research*, *105*(C10), 24971-24087.

Falqués, A., Montoto, A., & Vila, D. (1999). A note on hydrodynamic instabilities and horizontal circulation in the surf zone. *Journal of Geophysical Research*, *104*(C9), 20605-20615.

Felippa, C. A. (2004). *Introduction to Finite Element Methods* (Lecture Notes ed.). Department of Aerospace Engineering Sciences and Center for Aerospace Structures: University of Colorado Boulder, Colorado 80309-0429, USA.

Finlayson, D. (2005). Combined bathymetry and topography of the puget lowland, washington state,university of washington. *http://www.ocean.washington.edu/data/pugetsound/*.

FitzGerald, D. (1988). Shoreline erosional-depositional processes associated with tidal inlets. In D. Aubrey & L. Weishar (Eds.), *Hydrodynamics and sediment dynamics of tidal inlets: Lecture notes on coastal and estuarine studies* (p. 186-225). New York: Springer-Verlag, Inc.

Forester, C. (1979). Higher order monotonic convective difference schemes. *Journal of Computational Physics, 23*, 1-22.

Fredsøe, J. (1984). Turbulent boundary layer in wave-current interaction. *J. Hydraul. Eng, 110*, 1103 - 1120.

Fredsøe, J., & Deigaard, R. (1992). *Mechanics of coastal sediment transport advanced series on ocean engineering* (1st, Vol. 3 ed.). Singapore: World Scientific.

Gallagher, E., Elgar, S., & Thornton, E. (1998). Observations and predictions of megaripple migration in a natural surf zone. *Nature, 394*, 165-168.

Gao, s., & Zhu, D. (1988). The profile of jiangsus mud coast. *Journal of Nanjing University (Natural Science Series),In Chinese with English abstract, 21(1)*, 75-84.

Gelfenbaum, G., Mumford, T., Brennan, J., Case, H., Dethier, M., Fresh, K., ... Woodson, D. (2006). Coastal habitats in puget sound: A research plan in support of the puget sound nearshore partnership. *Puget Sound Nearshore Partnership Report No. 2006-1, U.S. Geological Survey, Seattle, Washington*.

Gershenfeld, N. (1999). *The nature of mathematical modeling*. Cambridge, UK: Cambridge University Press.

Ghisalberti, M., & Nepf, H. M. (2002). Mixing layers and coherent structures in vegetated aquatic flows. *Journal of Geophysical Research, 107(C2)*, 3-1 to 3-11.

Graf, W. (1971). *Hydraulics of sediment transport* (1st ed.). New York: McGraw-Hill.

Grasmeijer, B., & Van Rijn, L. (1998). Breaker bar formation and migration. *Proceedings of the 26th International Conference on Coastal Engineering, Copenhagen, Denmark. ASCE, New York*, 2750-2758.

Grizzle, R. E., Short, F. T., Newell, C. R., Hoven, H., & Kindblom, L. (1996). Hydrodynamically induced synchronous waving of seagrasses: monami and its possible effects on larval mussel settlement. *Journal of Experimental Marine Biology and Ecology, 206*, 165-177.

Groeneweg, J. (1999). Wave-current interactions in a generalised lagrangian mean formulation. *Dissertation of TU Delft University*, 154. (ISBN 90-9013073-X)

Grunnet, N. M., Walstra, D.-J. R., & B.G., R. (2004). Process-based modelling of a shoreface nourishment. *Coastal Engineering, 51*, 581-607.

Hanes, D., C.C., J., Thosten, E., & Vincent, C. (1998). Field observation of nearshore wave-seabed interactions. *Coastal Dynamics '97, ASCE*, 11-18.

Hasselmann, K., Barnett, T., Bouws, E., Carlson, H., Cartwright, D., Enke, K., ... Walden, H. (1973). Measurements of wind-wave growth and swell decay during the joint north sea wave project (JONSWAP). *Dtsch. Hydrogr. Z. Suppl., 12*, A8.

Hay, A. E., & Wilson, D. (1994). Rotary sidescan images of nearshore bedform evolution during a storm. *Marine Geology, 119*, 57-65.

Hervouet, J. (2007). Hydrodynamics of free surface flows: Modelling with the finite element method. *John Wiley & Sons, Ltd, ISBN 9780470035580*.

Hibma, A., de Vriend, H., & Stive, M. (2003). Numerical modelling of shoal pattern formation in well-mixed elongated estuaries. *Estuarine, Coastal and Shelf Science, 57*, 981-991.

Hibma, A., Stive, M., & Wang, Z. (2004). Estuarine morphodynamics. *Coastal Engineering, 51*, 765-778.

Holthuijsen, L. (2005). *Waves in oceanic and coastal waters.* Cambridge University Press.

Horikawa, K. (1988). *Nearshore dynamics and coastal processes: Theory, measurement, and predictive models* (1st ed.). Japan: University of Tokyo Press.

Huang, W., & Spaulding, M. (1996). Modelling horizontal diffusion with sigma coordinate system. *Journal of Hydraulic Engineering, 122 (6)*, 349-352.

Hudson, J., Damgaard, J., Dodd, N., Chesher, T., & Cooper, A. (2005). Numerical approaches for 1d morphodynamic modelling. *Coastal Engineering, 52*, 691-707.

Hulscher, S. (1996). Tidal-induced large-scale regular bed form patterns in a three-dimensional shallow water model. *J. Geophys. Res., 101*, 20727-20744.

Hulscher, S., De Swart, H., & De Vriend, H. (1993). The generation of offshore tidal sand banks and sand waves. *Cont. Shelf Res., 13*, 1183-1204.

Huthnance, J. (1982). On one mechanism forming linear sand banks. *Estuarine and Coastal and Shelf Science, 14*, 79-99.

Ikeda, S. (1982). Lateral bed-load transport on side slopes. *ASCE Journal of the Hydraulic Division, 108(11)*, 1369-1373.

Isobe, M., & Horikawa, K. (1982). Study on water particle velocities of shoaling and breaking waves. *Coastal Engineering in Japan, 25*.

Johnson, H. K., & Zyserman, J. A. (2002). Controlling spatial oscillations in bed level update schemes. *Coastal Engineering, 46*, 109-126.

Katopodi, I., & Ribberink, J. (1992). Quasi-3d modelling of suspended sediment transport by currents and waves. *Coastal Engineering, 18*, 83-110.

Ke, D., Healy, T., & Lu, S. (2009). Carrying capacity of red-crowned cranes in the national yancheng nature reserve, jiangsu, china. *International Journal of Ecology & Development, 12(WO9)*, 75-87.

Kernkamp, H., Van Dam, A., G.S., S., & de Goede, E. (2011). Efficient scheme for the shallow water equations on unstructured grids with application to the continental shelf. *Ocean Dynamics, 61*, 1175-1188.

Kingston, K. S., Ruessink, B. G., van Enckevort, I., & A., D. M. (2000). Artificial neural network correction of remotely sensed sandbar location. *Marine Geology, 169*, 137-160.

Klein, M. D. (2006). Modelling rhythmic morphology in the surf zone. *Ph.D. dissertation of TU Delft University*, 165. (ISBN 978-90-90201-85-6)

Klijn, F. (1997). A hierarchical approach to ecosystems and its implications for ecological land classification; with examples of ecoregions, ecodistricts and ecoseries of the netherlands. *Phd Dissertation of Leiden University*, 186.

Komar, P. D. (1998). *Beach processes and sedimentation* (2nd ed.). New York: Prentice Hall.

Kragtwijk, N., Zitman, T., Stive, M., & Wang, Z. (2004). Morphological response of tidal basins to human interventions. *Coastal Engineering, 21*, 207-221.

Kriebel, D., & Dean, R. (1985). Numerical simulation of time-dependent beach and dune erosion. *Coastal Engineering, 9*, 221-245.

Kubatko, E. J., Westerink, J. J., & Dawson Clint, N. (2006). An unstructured grid morphodynamic model with a discontinuous galerkin method for bed evolution. *Ocean Modelling, 15*, 71-89.

Larsen, L., Cannon, G., & Choi, B. (1985). East china sea tide currents. *Cont. Shelf Res., 4(1-2)*, 77-103.

Larson, M., Hanson, H., & Kraus, N. (1987). *Analytical solutions of the oneline model of shoreline change* (Technical Report No. X0267.10). Delft, the Netherlands: US Army Corps of Engineers.

Latteux, B. (1995). Techniques for long-term morphological simulation under tidal action. *Marine Geology, 126*, 129-141.

Leendertse, J. (1987). A three-dimensional alternating direction implicit model with iterative fourth order dissipative nonlinear advection terms. *WD-333-NETH, The Netherlands Rijkswaterstaat*.

Lesser, G. (2009). An approach to medium-term coastal morphological modelling. *CRC Press/Balkema, ISBN 978-0-415-55668-2*, 238 pp.

Lesser, G., Roelvink, J., Kester, J., & Stelling, G. (2004). Development and validation of a three-dimensional morphological model. *Coastal Engineering, 51*, 883-915.

Li, F., Dyt, C., & Griffiths, C. (2006). Multigrain sedimentation/erosion model based on cross-shore equilibrium sediment distribution: Application to nourishment design. *Estuarine, Coastal and Shelf Science, 67*, 664-672.

Li, H. (2009). Spatial pattern dynamics in aquatic ecosystem modeling. *Ph.D. Dissertation of Technische Universiteit Delft, UNESCO-IHE*.

Liu, Z., & Xia, D. (1983). A preliminary study on tidal current ridges. *Oceanologia et Limnologia Sinica (in Chinese), 14(3)*, 286-295.

Long, W., Kirby, J. T., & Shao, Z. (2008). A numerical scheme for morphological bed level calculations. *Coastal Engineering, 55*, 167-180.

Lopez, F., & Garcia, M. H. (1998). Open-channel flow through simulated vegetation: Suspended sediment transport modelling. *Water Resources Research, 34(9)*, 2341-2352.

Lu, P. (2005). Scaled model for tidal current and sediment transport in lng port in rudong, jiangsu. *Nanjing Hydraulic Research Institute, Report, No. R05115, R05115*.

Marchuk, G., & Kagan, B. (1984). *Ocean tides: Mathematical models and numerical experiments*. Cambridge, UK: Pergamon Press, ISBN 0-08-026236.

May, R. (1976). Simple mathematical models with very complicated dynamics. *Nature, 261*, 459, doi:10.1038/261459a0.

Mei, C. (1983). *The applied dynamics of ocean surface waves*. Wiley, New York, 740. pp.

Mellor, G., & Blumberg, A. (1985). Modelling vertical and horizontal diffusivities and the sigma coordinate system. *Monthly Weather Review, 113*, 1379-1383.

Mendez, F., & Losada, I. (2004). An empirical model to estimate the propagation of random breaking and nonbreaking waves over vegetation fields. *Coastal Engineering, 51*, 103-118.

Meulé, S., PINAZO, C., DEGIOVANNI, C., BARUSSEAU, J.-P., & LIBES, M. (2001). Numerical study of sedimentary impact of a storm on a sand beach simulated by hydrodynamic and sedimentary models. *OCEANOLOGICA ACTA, 24*, 417-424.

Meyer-Peter, E., & Mueller, R. (1948). Formulas for bed-load transport. *Proc., 2nd 1nt. IAHR Congress, Stockholm, Sweden*, 39-64.

Modelling storm impacts on beaches, dunes and barrier islands. (2009). *Coastal Engineering, 56*(11-12), 1133 - 1152.

Monden, M. (2010). Modeling the interaction between morphodynamics and vegetation in the nisqually river estuary. *Msc Thesis, TU Delft, the Netherlands*.

Morphodynamic modeling using the telemac finite-element system. (2011). *Computers & Geosciences*(0), -.

Murray, A. (2003). Contrasting the goals, strategies and predictions associated with simplified numerical models and detailed simulations. In P. Wilcock & R. Iverson (Eds.), *Prediction in geomorphology, agu, geophysical monograph 135*.

Murray, A. (2007). Reducing model complexity for explana-tion and prediction. *Geomorphology, 90*.

Murray, A., Knaapen, M., Tal, M., & Kirwan, M. L. (2008). Biomorphodynamics: Physical-biological feedbacks that shape landscapes. *Water Resour. Res., 44*, W11301.

Nicholson, J., Broker, I., Roelvink, J., Price, D., Tanguy, J., & Moreno, L. (1997). Intercomparison of coastal area morphodynamics models. *Coastal Engineering, 31*, 97-123.

Nielsen, P. (1992). *Coastal bottom boundary layers and sediment transport* (1st ed.). Singapore: World Science.

Nisqually River Basin Plan. (2008). Nisqually river basin plan. *Pierce County Public Works and Utilities Water Programs Division*.

Pan, S., MacDonald, N., Williams, J., OConnor, B., Nicholson, J., & Davies, A. (2007). Modelling the hydrodynamics of offshore sandbanks. *Continental Shelf Research*, *27*, 1264-1286.

Pape, L., Ruessink, B. G., Wiering, M. A., & Turner, I. L. (2007). Recurrent neural network modeling of nearshore sandbar behavior. *Neural Networks*, *4*, 1-22.

Parker, G., Lanfredi, N., & P., S. D. J. (1982). Seafloor response to flow in a southern hemisphere sand-ridge field: Argentina inner shelf. *Sedimentary Geology*, *33*, 195-216.

Partheniades, E. (1965). Erosion and deposition of cohesive soils. *ASCE Journal of Hydraulic Division*, *91(HY1)*, 105-139.

Partheniades, E. (2009). *Cohesive sediments in open channels: properties, transport, and applications.* Butterworth-Heinemann.

Pelnard, C. (1956). Essai de thorie de l'volution des formes de rivage en plages de sable et de galets. *Soc. Hydrotechnique de France, Quatimes Journes de l'Hydraulique. Les nergies de la Mer, Tome I, Question III*, 289-298.

Phillips, J. (1995). Biogeomorphology and landscape evolu-tion: The problem of scale. *Geomorphology*, *13*, 337-347.

Postma, L., & Hervouet, J.-m. (2007). Compatibility between finite volumes and finite elements using solutions of shallow water equations for substance transport. *International Journal for Numerical Methods in Fluids*, 53, 1495-1507.

Puget Sound Partnership. (2008). Nisqually watershed chinook salmon recovery plan 3 year workprogram 2008-2010. *Puget Sound Partnership*.

Rakha, K. (1998). A quasi-3d phase-resolving hydrodynamic and sediment transport model. *Coastal Engineering*, *34*, 277-311.

Ren, M. (1986). Tidal mud flat. *In: Ren, M.E. (Ed.), Modern sedimentation in the coastal and nearshore zones of China. China Ocean Press, Beijing.*

Ribberink, J. (1987). Mathematical modelling of one-dimensional morphological changes in rivers with non-uniform sediment. *Ph.D. dissertation, Delft University of Technology.*

Ris, R., Holthuijsen, L., & Booij, N. (1999). A third generation wave model for coastal regions: 2. verification. *J. Geophys. Res.*, *104*, 7667- 7681.

Roelvink, J. (2006). Coastal morphodynamic evolution techniques. *Coastal Engineering*, *53*, 277-287.

Roelvink, J., & Brøker, I. (1993). Cross-shore profile models. *Coastal Eng.*, *21*, 163-191.

Roelvink, J., & Reniers, A. (2012). A guide to modelling coastal morphology. *World Scientific Publishing Co Pte Ltd, Advances in Coastal and Ocean Engineering Series*, ISBN 9789814304252.

Rózyński, G. (2003). Data-driven modeling of multiple longshore bars and their interactions. *Coastal Engineering*, *48*, 151-170.

Ruessink, B. (2005). Predictive uncertainty of a nearshore bed evolution model. *Continental Shelf Research*, *25*, 1053-1069.

Sanz, L. d. l. P. (1999). Variables aggregation in a time discrete linear model. *Mathematical Biosciences*, *157(1-2)*, 91-109.

Sanz, L. d. l. P. (2000). Time scales in stochastic multiregional models. *Nonlinear Analysis:Real World Applications*, *1*, 89-122.

Schanz, A., & Asmus, H. (2003). Impact of hydrodynamics on development and morphology of intertidal seagrasses in the wadden sea. *Mar. Ecol. Prog. Ser.*, *261*, 123-134, doi:10.3354/meps261123.

Schoonees, J., & Theron, A. (1995). Evaluation of 10 cross-shore sediment transport/morphological models. *Coastal Engineering*, *25*, 1-41.

Scott, T., & Mason, D. (2007). Data assimilation for a coastal area morphodynamic model: Morecambe bay. *Coastal Engineering*, *54*, 91-109.

Seminara, G. (1998). Stability and morphodynamics. *Meccanica*, *33*(Number 1, February 1998), 59-99.

Sleath, J. (1984). *Sea bed mechanics* (1st ed.). New York: Wiley.

Smit, M., Aarninkhof, S., Wijnberg, K., González, M., Kingston, K., Southgate, H., ... R., M. (2007). The role of video imagery in predicting daily to monthly coastal evolution. *Coastal Engineering*, *54*, 539-553.

Smith, G. (1985). *Numerical solution of partial differential equations: Finite difference methods* (3rd ed.). Clarendon Press, Oxford, 337 pp.

Soulsby, R. (1997). *Dynamics of marine sands* (1st ed.). London, U.K.: Thomas Telford.

Spielmann, K., Astruc, D., & Olivier, T. (2004). Analysis of some key parametrizations in a beach profile morphodynamical model. *Coastal Engineering*, *51*, 1021-1049.

Stelling, G. (1984). On the construction of computational methods for shallow water flow problem. *Rijkswaterstaat Communications vol. 35. Government Printing Office, The Hague, The Netherlands*, 224.

Stelling, G., & Duinmeijer, S. (2003). A staggered conservative scheme for every froude number in rapidly varied shallow water flows. *International Journal for Numerical Methods in Fluids*, *43*, 1329-1354.

Stelling, G., & Leendertse, J. (1991). Approximation of convective processes by cyclic adi methods. *Proceeding of the 2nd ASCE Conference on Estuarine and Coastal Modelling, Tampa, ASCE*, 771-782.

Stelling, G., & Van Kester, J. (1994). On the approximation of horizontal gradients in sigma coordinates for bathymetry with steep bottom slopes. *International Journal for Numerical Methods in Fluids*, *18*, 915-955.

Stive, M., & de Vriend, H. (1995). Modelling shoreface profile evolution. *Marine Geology*, *126*, 235-248.

Stive, M., & Wang, Z. (2003). Morphodynamic modelling for tidal basins and coastal inlets. , 367-392.

Stoker, J. (1957). *Water waves; the mathematical theory with applications.* 567 blz.: New York Interscience.

Sutherland, J., Peet, A., & Soulsby, R. (2004). Evaluating the performance of morphological models. *Coastal Eng.*, *51*, 917-939.

Suzuki, T. (2011). Wave dissipation over vegetation fields. *Ph.D. dissertation of TU Delft University*, 194.

Suzuki, T., Marcel, Z., Bastiaan, B., Martijn, M., & Siddharth, N. (2011). Wave dissipation by vegetation with layer schematization in swan. *Coastal Engineering*, *51*, 103-118.

Svendsen, I. (2002). *Introduction to nearshore hydrodynamics.* World Scientific Pub Co Inc.

Swift, D. J. P., Parker, G., Lanfredi, N. W., & Figge, G. P. K. (1978). Shoreface-connected sand ridges on american and european shelves: A comparison. *Estuarine Coastal Mar. Sci.*, *7*, 257-273.

Szmytkiewicz, M., Biegowski, J., Kaczmarek, L. M., Okrój, Ostrowski, T. R., Pruszak, Z., ... Skaja, M. (2000). Coastline changes nearby harbour structures: comparative analysis of one-line models versus field data. *Coastal Engineering*, *40*, 119-139.

Tang, G., Shafer, S. L., Bartlein, P. J., & Holman, J. O. (2009). Effects of experimental protocol on global vegetation model accuracy: A comparison of simulated and observed vegetation patterns for asia. *Ecological modelling*, *220*, 1481-1491.

Temmerman, S., Bouma, T., de Koppel, J. V., der Wal, D. V., Vries, M. D., & Herman, P. (2007). Vegetation causes channel erosion in a tidal landscape. *Geology*, *35*, 631-634.

Temmerman, S., Bouma, T., Vries, M. D., Wang, Z., Govers, G., & Herman, P. (2005). Impact of vegetation on flow routing and sedimentation patterns: three-dimensional modeling for a tidal marsh. *Journal of Geophysical Research*, *110*, F04019.

Thompson, S., G., K., & McMahon, S. M. (2008). Role of biomass spread in vegetation pattern formation within arid ecosystems. *Water Resour. Res.*, *44*, W10421, doi:10.1029/2008WR006916.

Thornton, E. B., Swayne, J., & J.R., D. (1998). Small-scale morphology across the surf zone. *Marine Geology*, *145*, 173-196.

Trevethan, M., Hubert, C., & Richard, B. (2008). Turbulence and turbulent flux events in a small subtropical estuary. *Report CH65/07*, ISBN-9781864998993.

Trowbridge, J. H. (1995). A mechanism for the formation and maintenance of shore-oblique sand ridges on storm-dominated shelves. *J. Geophys. Res.*, *100 (C8)*, 16071-16086.

Tsoularis, A. (2001). Analysis of logistic growth models. *Res. Lett. Inf. Math. Sci*, *2*, 23-46.

Tung, T., Walstra, D., Graaff, J. v. d., & Stive, M. (2009). Morphological modeling of tidal inlet migration and closure. *Journal of Coastal Research*, *ICS2009*, 1080-1084.

Uittenbogaard, R. (2003). Modeling turbulence in vegetated aquatic flows. In *International workshop on riparian forest vegetated channels: hydraulic, morphological and ecological aspects.* Trento, Italy.

US Fish and Wildlife Service. (2005). Nisqually national wildlife refuge final comprehensive conservation plan. *US Fish and Wildlife Service.*

Van Den Burg, M. (1999). Charophyte colonization in shallow lakes. *Ph.D. dissertation of Vrij University Amsterdam*, 136.

Van De Meene, J. W. H., & Van Rijn, L. (2000). The shoreface-connected ridges along the central dutch coast part 1: Field observations. *Continental Shelf Research*, *20*, 2295-2323.

Van De Wegen, M. (2010). Modeling morphodynamic evolution in alluvial estuaries. *Ph.D. dissertation of TU Delft University.*

Van De Wegen, M., Wang, Z. B., Savenije, H. H. G., & Roelvink, J. A. (2008). Long-term morphodynamic evolution and energy dissipation in a coastal plain, tidal embayment. *J. Geophys. Res.*, *113*, F03001, doi:10.1029/2007JF000898.

van Dongeren, A., Plant, N., Cohen, A., Roelvink, D., Haller, M. C., & Cataln, P. (2008). Beach wizard: Nearshore bathymetry estimation through assimilation of model computations and remote observations. *Coastal Engineering*, *55*(12), 1016 - 1027. doi: DOI:10.1016/j.coastaleng.2008.04.011

van Duin, M., Wiersma, N., Walstra, D., & M.J.F., S. (2004). Nourishing the shoreface: observations and hindcasting of the egmond case, the netherlands. *Coastal Engineering*, *51*, 813-837.

Van Ledden, M. (2003). Sand-mud segregation in estuaries and tidal basins. *Dissertation of TU Delft University*, 221. (ISBN 90-9016786-2)

Van Leeuwen, S., N., D., Calvete, D., & A., F. (2004). Linear evolution of a shoreface nourishment. *Coastal Engineering*, *54*, 417-431.

Van Leeuwen, S., van der Vegt, M., & de Swart, H. (2003). Morphodynamics of ebb-tidal deltas: a model approach. *Estuarine, Coastal and Shelf Science*, *57*, 899-907.

Vanoni, V. (1975). *Sedimentation engineering* (1st ed.). New York: ASCE.

Van Rijn, L. (1987). *Mathematical modeling of morphological processes in the case of suspended sediment transport* (Communication No. 382). Delft, the Netherlands.

Van Rijn, L. (1993). *Principles of sediment transport in rivers, estuaries and coastal seas* (1st ed.). Amsterdam: Aqua Publications.

Van Rijn, L. (2001). Approximation formulae for sand transport by currents and waves and implementation in delft-mor [Technical Report]. (Z3054.20).

Van Rijn, L. (2006). *Principles of sediment transport in rivers, estuaries, and coastal seas* (update 2006 ed.). Blokzijl, The Netherlands, www.aquapublications.nl: Aqua Publications.

Van Rijn, L. (2007a). Unified view of sediment transport by currents and waves. 1: Initiation of motion, bed roughness and bed-load transport. *J. Hydraul. Eng.*, *133(6)*, 649-667.

Van Rijn, L. (2007b). Unified view of sediment transport by currents and waves. 2: Suspended transport. *J. Hydraul. Eng.*, *133(6)*, 668-689.

Van Rijn, L. (2007c). Unified view of sediment transport by currents and waves. 3: Graded beds. *J. Hydraul. Eng.*, *134(7)*, 761-775.

Van Rijn, L. (2007d). Unified view of sediment transport by currents and waves. 4: Application of morphodynamic model. *J. Hydraul. Eng.*, *134(7)*, 776-793.

Van Rijn, L., & Walstra, D. (2003, Oct.). Modelling of sand transport in delft3d-online [Technical Report]. (Z3624, final).

Van Rijn, L., Walstra, D., Grasmeijer, B., & Sutherland, J. (2003). The ctability of cross-shore bed evolution of sandy beaches at the time scale of storms and seasons using process-based profile models. *Coastal Engineering*, *47*, 295-327.

Van Rijn, L., Walstra, D., & Van Ormondt, M. (2004). Description of transpor2004 and implementation in delft3d-online [Technical Report]. (Z3748.10).

Vila-Concejo, A., Ferreira, O., Matias, A., & Dias, J. (2003). The first two years of an inlet: sedimentary dynamics. *Continental Shelf Research*, *23*, 1425-1445.

Viles, H. A. (1988). *Biogeomorphology*. London, Blackwell: Oxford, ISBN: 0-631-15405-1.

Villaret, C., & Gonzales, M., DE LINARES. (2005). Sisyphe release 5.5 validation manual. , *EDF R&D*, pp.63.

Vittori, G., & Blondeaux, P. (1992). Sand ripples under sea waves, part 3. brick-pattern ripples formation. *J. Fluid Mech.*, *239*, 23-45.

Vreugdenhil, C. (1994). *Numerical methods for shallow-water flow*. 261 blz.: Dordrecht Kluwer.

Walgreen, M., Calvete, D., & De Swart, H. (2002). Growth of large-scale bed forms due to storm-driven and tidal currents: a model approach. *Cont.Shelf Res.*, *22*, 2777-2793.

Walstra, D., Roelvink, J., & Groeneweg, J. (2000). Calculation of wave-driven currents in a 3D mean flow model. In *Proceedings of the 27th international conference on coastal engineering, Sydney* (Vol. 2, p. 1050-1063).

Wang, Z. (1992). Theoretical analysis on depth-integrated modelling of suspended sediment transport. *Journal of hydraulic research*, Vol 30, No.3.

Wang, Z., Louters, T., & De Vriend, H. (1995). Morphodynamic modelling for a tidal inlet in the waden sea. *Marine Geology*, *126*, 289-300.

Wang-Y, Y. (2002). Radiative sandy ridge field on continental shelf of the yellow sea (in chinese). *China Environmental Science Press*.

Wang-Y, Y., & Zhu, D. (1998). Sedimentation chrematistics and evolution of radial shape sand ridge group at southern yellow sea, china. *Science in China (series D)*, *28(5)*, 386-393.

Wenneker, I. (2003, April). *Solution of the shallow water equations using unstructured grids* (Technical Report No. X0267.10). Delft, the Netherlands: WL|Delft Hydraulics.

Werner, B. (2003). Modeling landforms as self-organized, hierarchical dynamical systems. In P. Wilcock & R. Iverson (Eds.), *Prediction in geomorphology, agu, geophysical monograph 135*.

Westerink, J., Luettich, R., Feyen, J., Atkinson, J., Dawson, C., & Roberts, H. (2008). A basin to channel scale unstructured grid hurricane storm surge model applied to southern louisiana. *Monthly Weather Review*, *136(3)*, 833-864.

Whitham, G. (1974). *Linear and nonlinear waves*. New York: Wiley Interscience, 636. pp.

Wilcock, R., P.R.and Iverson. (2003). Prediction in geomorphology. In P. Wilcock & R. Iverson (Eds.), *Prediction in geomorphology, agu, geophysical monograph 135.*

Winterwerp, J., Uittenbogaard, R., van Kesteren, W., & Wang, Z. (1997). *Dynastar generic mud transport model : physical design study* (Technical Report No. 152 p.) Delft, the Netherlands: WL|Delft Hydraulics.

Woo, I., Turner, K., & Takekawa, J. (2010). Pre-restoration vegetation summary in the nisqually delta, fall 2009. *Unpublished data summary update, USGS, Western Ecological Research Center, San Francisco Bay Estuary Field Station, Vallejo, CA..*

Wright, L., & Short, A. (1984). Morphodynamic variability of surf zones and beaches: a synthesis. *Mar. Geol., 26*, 93-118.

Yalin, M. (1977). *Mechanics of sediment transport* (1st ed.). Oxford, U.K.: Pergamon.

Yan, Y., Song, Z., Xue, H., & Mao, L. (1999). Hydromechanics for the formation and development of radial sandbanks 2: Vertical characteristics of tidal flow. *Science in China (series D), 42(1)*, 22-29.

Yan, Y., Zhu, Y., & Xue, H. (1999). Hydromechanics for the formation and development of radial sandbanks 1: Plane characteristics of tidal flow. *Science in China (series D), 42(1)*, 13-21.

Yanagi, T., Morimoto, A., & Ichikawa, K. (1997). Co-tidal and co-range charts for the east china sea and the yellow sea derived from satellite altimetric data. *Journal of Oceanography, 53*, 303-309.

Yang, C. (1985). On the origin of jianggang radial sand ridges in yellow sea (in chinese). *Marine Geology and Quaternary Geology, 5(3)*, 35-44.

Ye, Q., Roelvink, J., Jagers, H. R. A., Postma, L., & van Beek, J. K. L. (2009). Coastal bio-geomorphology modeling: linking processes and scales. *Proceedings of River, Coastal and Esturary morphodynamics, Santa Fe, Argentina*, 559-566.

Zhang, D., & Zhang, J. (1996). M2 tidal wave in the radial shape sand ridges area (in chinese). *Journal of Hohai university, 24(5)*, 35-40.

Zhang, D., Zhang, J., & Zhang, C. (1998). Tide forms it - storm destroy it -tide recover it, a preliminary study on the mechanism of generation and evolution of the radial shape sand ridges in yellow sea. *Science in China (series D), 28(5)*, 394-402.

Zhang-C.K., & Wang, Y. (2009). Tidal flat reclaimation planning on the coast of jiangsu province, china. *Hohai University, Research report*, pp. 156.

Zhang-RS, & Chen, C. (1992). The evolution of the sand ridges along jiangsu coast and the possibility of tiaozi sand merging to mainland. *1992, Ocean Science Press.*

Zhang-RS, Shen, Y., LU, L., Yan, S., Wang, Y., Li, J., & Zhang, Z. (2004). Formation of spartina alterniflora salt marshes on the coast of jiangsu province, china. *Ecological engineering, 23*, 95-105.

Zhu, Y. (2003). Hydromechanical characteristics of the radial sandbanks in the southern china sea. *Ph.D. dissertation, Hohai University, China.*

Zhu, Y., Li, C., & Chang, R. (1995). Numerical simulation on generation of the radial shape sand ridges close to jianggang, and their paleoenvironmental values. *Journal of Tongji University, 23*, 226-230.

Zimmerman, J. (1981). Dynamics, diffusion and geomorphological significance of tidal residual eddies. *Nature, 290*, 549-555.

Zuo, P., Zhao, S., Liu, C., Wang, C., Teng, H., Zou, X., & Chen, H. (2009). Biodiversity conservation and landscape changes during 1970-2007 in the core area of yancheng biosphere reserve, jiangsu province, china. *Proceedings of the 11th International Conference on Environmental Science and Technology, Chania, Crete, Greece, 3-5 September 2009*, B-1033-1040.

Acknowledgements

My sincere gratitude goes first to my promoter Prof.dr.ir. (Dano) J.A. Roelvink for giving me opportunity to carry out this PhD research. His invaluable guidance and inspiring discussions during the last years contribute a lot to this work. Every discussion with him is informative to me. His patience and trust on me are always my strongest support. Constant encouragement and stimulations from Dr. (Bert) H.R.A. Jagers, as both a daily supervisor and a friend, is greatly appreciated. I am grateful to Prof. Arthur Mynett for his stimulating discussions and suggestions. He is always so kind to involve me in his students group and his office/coffee corner was so cozy, and was arranged as a MIT-style corner for us with all kinds of ideas and information. It was Prof. Roland Price who picked me up from China in 2004 and gave me the opportunity to Delft for a M.Sc to learn Hydroinformatics. And his lectures on hydrodynamics and flood routing always fascinate me.

Deltares software center and the Netherlands Ministry of Infrastructure and Environment are appreciated for their financial support to this study.

Leo Postma, Jan van Beek are appreciated for the regular discussions and guidances in applying and programming Delft3D WAQ. This work could not be done without their constant support.

I am indebted to Dano and Bert for their efforts in correcting and improving my poorly written manuscripts. I want to thank my other committee members, Prof. Lee, Prof. Mynett, Prof. Stelling, Prof. Stive and especially Prof. Herman for their time in reading, helpful discussions, critical comments, and suggestions on my research and on this thesis.

I am so grateful to the colleagues from Deltares software center, such as, Frank Hoozemans, Arthur Bart, Jan van Kester, Maart Bosboom, Adri Mourits, Herman Kernekamp, Rinie Hoff (forgive me that I cannot list all of you, otherwise I just print the whole smoelboek out). Thanks to your patience and hospitality to guide me in the past years. I regarded myself so lucky to be with you during this scientific adventure. Your support and help encouraged me during my study.

I am so grateful to the colleagues from Hydraulic Engineering Unit of Deltares, Klaas Jan Bos, Wiel Tilmans, JJ, Cilia, Hans, Robin, Radha, etc. I learnt quite a lot from all of you on carrying out a consultancy work, even though to my dreaming of PE, the time is too short and the lessons learnt are still far from enough. Special thanks to Arjen Luijendijk that he kindly involved me in the Build with Nature projects and kept stimulating my ability of networking, which I am lack of.

I am so grateful to the colleagues from ZKS of Deltares, Prof. Wang Zhengbin, Maarten van Ormondt, PK Tonnon, Jamie and others. Time with you guys is always relaxed and full of fun.

I would like to thank Megena, Sanjay, Anke, Kees Sloff, Eric Mosselman from the river morphology group of Deltares. Thanks for your support and all the interesting discussions.

Mick van der Wegen, Giles Lesser, Dissa, Ali, Guo, Wan and other colleagues, PhD fellows in CEPD of UNESCO-IHE are appreciated for all the discussion and encouragement.

Guy Gelfenbaum and Andrew Stevens from USGS are appreciated for their permission to use the data of Nisqually estuary, Puget Sound. Rijkswaterstaat and Ellis Penning from Deltares are appreciated for their permission to use the vegetation data of the Lake Veluwe. Martijn Monden is thankful for his help on the modeling efforts about Puget Sound.

I would like to thank Mindert de Vries from Deltares, Stijn Temmerman from University of Antwerp, Johan van Koppel, Tjerd Bouma from NIOO. Thanks to you all for your evaluable inputs into this study from ecological point of view, which I still know very little.

I would like to thank Prof. Solomatine, Biswa, Dr. Zhou, Jan Luijendijk, Jeltsje, Klaas, Peter, Lydia, Prof. Uhlenbrook and other our ROPA-Run team runners. We had so much fun and good times together. I miss all of you. Enormous thanks to Jolanda, Maria, Sylvia and colleagues from the back office for your consistent supports in the past years. Enormous thanks to Anique Alaoui-Karsten and Peter Stroo, without your help, this book cannot be printed out in such a short time.

Thanks to the Chinese community in Delft, I cherished the time with you all. I would like to list some of them: Li Hong, Wang Wen, Xu Min, Bai Yuqian, Qi Hui, Xu Zhuo, Zhu Xuan, Li Shengyang, Yang Zhi, Li Rongchao and Omar, Wang Zhuoyu, Sun Tao and Wang Yuan, Xu Wei, Wu Xi, Wu Xiaomin,Jiang Huaying, Liang Xiao, Huang Mingxin, Tu Huizhao, Li Hao, Cui Haiyang, An Ran. Lin Yuqing. Chen Qiuwen. Wang Yubin, Ma Ying, Chen Hui, Chu Ao, Peng Zhong, Sun Tong, Young Wang Li in Antwerp, and so forth and forth. I wish you all the best!

Feng Han and Chen Qing are appreciated for their help to design the cover of this book.

Special thanks to Johanna Boersen who help me on the first shoot on the samenvatting. Let us see, she will probably be a qualified morphologist soon.

Thanks to Prof. Yu Guohua, Prof. Lu Peidong, Prof. Chen Yongping, Dr. Wang Yanhong, Dr. Lu Qimiao, Dr. Chen Yong, Dr. Zou Shan, Dr. Chen Kefeng, Xu Zhuo and so on from Nanjing Hydraulic Research Institute, China, and Prof. Zhou Qiwu, Sun Zhilin, Qiu Jianli, Zhu Yongkang, Zhou Dacheng, Wang Zhongtao from Zhejiang University. Thanks for your enoumous help and guide me to explore the coastal engineering profession. Prof. Pan Chenghong, Prof.Yao Shengchu and Ms. Li Yuqing are grateful for your support on my way towards computational hydraulics and scientific programming. You are always be remembered by me.

Thanks to Dr. Pan Shunqi from University of Plymouth for your understanding and support during this study.

Thanks to Gao Min. We have shared so much good time together. I really miss you, buddy. I am grateful to Teng Ling who gave me so kind help during my initial period to Delft.

Thanks to Dr. Wang Li and Zhao Hongli. A friend in need is a friend indeed. You are always my support in soul.

Ken, my dear brother, I am so lucky to meet you! How is it possible that exists another stubborn person, exactly like me, on the earth? I always appreciate your criticism and care, which helps me grow mentally.

Thanks to Ye Danxia, Fan Wei and little Aurora. Thanks to my parents, Ye Jiacai and Mei Xianglan. Pls excuse my absence when you missed me for so long in time and so far in distance.

The last but not the least, my thanks to Taoping and Ivy, forgive my absent mind and my bad temper during this long journey.

Delft, June 2012.

About the Author

Qinghua YE was born in Wuxue, Hubei of China on January 10th of 1977. He studied coastal engineering at Zhejiang University (former Hangzhou University) from 1994 to 1998 and graduated with first class honours (cum laude). He started in 1998 and graduated at 2001 as an Msc in specialization of coastal engineering and sediment transport in Nanjing Hydraulic Research Institute. From 2001 to 2004, he worked as an advisor and researcher in the section of River and Harbour department in Nanjing Hydraulic Research Institute, where he was involved with tens of projects on coastal and estuarine dynamics, including the design, construction of harbours and navigation channels, land reclamation, environment assessment of intake and outfall for thermal plant power stations, etc.

In 2004 he travelled to Delft, the Netherlands to study Hydroinformatics in UNESCO-IHE and was awarded a 2nd Msc in 2006. His thesis project is on the morphodynamic modelling in the Haringvliet.

In 2006, he started to persue his phd in Delft Hydraulics (now named Deltares). From September of 2010 till now he was full time employed by Deltares as a consultant and researcher, first in the Hydraulic Engineering unit and later in the Deltares software centre.

He is also a Microsoft certified software developer in C++ and C#, a Sun certified Java programmer and a Cisco certified network associate.

He married with Taoping Wan in 2002 and got a daughter, Ivy Zixin Ye in 2008.